Construction – The Third Way

Construction – The Third Way

Managing Cooperation and Competition in Construction

John Bennett

OXFORD AUCKLAND BOSTON JOHANNESBURG MELBOURNE NEW DELHI

Butterworth-Heinemann
Linacre House, Jordan Hill, Oxford OX2 8DP
225 Wildwood Avenue, Woburn, MA 01801-2041
A division of Reed Educational and Professional Publishing Ltd

A member of the Reed Elsevier plc group

First published 2000

British Library Cataloguing in Publication Data
Bennett, John
Construction : the third way
1. Construction industry – Great Britain – Management
I. Title
628'.068

ISBN 0 7506 3093 0

Library of Congress Cataloguing in Publication Data
A catalogue record for this book is available from the Library of Congress

Composition by Genesis Typesetting, Rochester, Kent
Printed and bound in Great Britain

Contents

Preface

The leading edge of the UK construction industry has changed focus in the 1990s. This is largely in response to demands from major customers for better value, delivered faster and more reliably. Developments in information technology have added to the pressure for change. The need for change is well documented in the Latham and Egan Reports (Latham, 1994; Construction Task Force, 1998). Latham recognized the need for the industry to move from its traditional adversarial approach to one based on cooperation and trust. Egan took this conclusion further in recommending an action plan that draws on the ideas of lean thinking and partnering. Both reports indicate the extent of the problems by suggesting tough targets for the industry to improve its performance year on year.

The leading edge of the industry has begun to deliver the kind of improvements that Latham and Egan demanded. My evidence for the existence of these improvements comes from research carried out as Director of The University of Reading's Centre for Strategic Studies in Construction. Between 1994 and 1998, with several research assistants, I undertook over 200 case studies of leading practice in the UK construction industry. Most of the case studies relate to building projects which traditionally give rise to fragmented and complicated organizations characterized by all the problems described in the Latham and Egan reports. The practical lessons from these case studies and other related research are described in two reports by Bennett and Jayes (1995, 1998) describing best practice partnering, and one by Bennett *et al.* (1996) which describes a blueprint for integrating design and construction processes in the UK construction industry.

The most significant of the data describing the improvements in the performance of the UK construction industry are given in *Table 1.3*. This table shows that leading practice is already achieving cost reductions of up to 50 per cent and time reductions of up to 80 per cent compared with traditional approaches.

The case studies were undertaken in cooperation with industry and it is good to have this opportunity to acknowledge the huge contributions of time and thought given by the many leading practitioners referred to in the published reports. The case studies form an important part of research which grew out of my earlier studies of management in the Japanese construction industry. A key stage of this was research undertaken in 1991 as Professor at the University of Tokyo into the management methods used by the top five Japanese construction firms. My ideas were further influenced by working as the lead academic in research to provide a basis for Europe's policy towards construction for the European Union (Atkins *et al.*, 1994).

The European study served to bring a number of ideas together as the research team produced structured descriptions of the construction sector of all European Union countries, Japan and the USA. One part of this work produced comparisons of the relative efficiency of these major construction industries. The results in *Table 1.1* are supported by other similar comparisons which, taken together, provide two important insights. First, the results suggest that the US building industry's reputation for low cost production comes from its use of standard components and low safety and comfort levels, rather than being evidence of an efficient industry. Second, the results show that, comparing like with like, Japan has the most efficient building industry. My own research suggests that this superior performance results from an unusually integrated approach led by design build contractors and is based on a culture that supports cooperative, long-term relationships between firms.

All of this led fairly directly into research, funded by the Engineering and Physical Sciences Research Council (EPSRC), designed to understand the effects of cooperative behaviour in the UK's highly competitive construction industry. It focused particularly on the management actions needed to make long-term relationships effective. The early results from this EPSRC-funded research persuaded the Reading Construction Forum to commission two reports into best practice partnering (Bennett and Jayes, 1995, 1998), and what has now become the Design Build Foundation to commission a report into design build (Bennett *et al.*, 1996).

Bennett and Jayes (1998) provide a model (produced with the Reading Construction Forum's Partnering Task Force, ably chaired by Charles Johnson of Sainsbury's) of how industry and academia should work together. Over 12 months in 1997 and 1998 the case studies of partnering that I had done with the assistance of Sarah Jayes (now Sarah Peace) were reviewed at a series of intensive workshops designed to understand what was happening in practice. The workshops linked academic research and wide practical experience in an incredibly creative way as the Task Force of very experienced practitioners worked with Sarah and me to find consistent patterns in the case study material. The result, the model described by Bennett and Jayes (1998), is already widely used by customers and firms in the UK construction industry to guide their use of partnering.

The case studies are significant because they show that parts of the UK building industry have begun to use cooperative behaviour and that this provides substantial benefits. However, these improvements have shaky foundations. It would be easy for those involved to slip back into traditional attitudes and methods in response, for example, to a downturn in demand. The need for the fundamental change in culture called for by the Egan Report remains. Change of this kind requires a paradigm shift, which is why Sir John Egan called his report *Rethinking Construction* (Construction Task Force, 1998). That is exactly what is required: the industry needs to think about its work in a fundamentally different way. It needs to see its customers, the communities it serves, the various parts of the industry and the relationships between them differently. It needs what is properly called a paradigm shift.

That is the background for this book which grew out of a request from publishers, Butterworth-Heinemann, that I should edit my earlier book (Bennett, 1991), so that it could be re-published. This earlier book describes a general theory of construction project management and illustrates practical implications with examples drawn from international best practice as it was in 1990. The book was based on research into leading practice in the UK, Western Europe, USA and Japan. In response to Butterworth-Heinemann's request, I made several attempts to edit the earlier book. Slowly and painfully it became clear that the construction industry, at least the leading edge of it in the UK, has changed and that the 1990 theory and practice no longer provide adequate descriptions. Any editing of the 1990 descriptions would be inadequate and so a fundamentally different book is needed to describe my current understanding of theory and practice.

This realization coincided with my decision to resign from the Directorship of the Centre of Strategic Studies in Construction to give myself time to think carefully about the changes taking place in construction practice. As a result, since 1997, I have had time to read about and discuss changes in other industries and in scientific thinking that have important similarities to what is happening in construction. These larger developments helped me to understand the significance of my case studies.

The resulting ideas are described in this book which, inevitably, is just one stage of a journey that began in what I now regard as an outdated paradigm based on an elemental, hierarchical view of the world in which progress is achieved by top-down management decisions. The journey includes research into Japanese management which, for most of the last 20 years, has out-performed the West in key major industries. The main differences in their approach centre on the use of long-term cooperative relationships that encourage workers at all levels to search for continuous improvements in performance. In response to the Japanese challenge, leading manufacturing firms in the USA developed partnering as a way of introducing cooperative behaviour into a culture dominated by an unquestioning faith in competitive market forces. Partnering emerged first in manufacturing and was subsequently applied to construction, initially in the USA and then in many other countries including the UK. In researching these developments it became clear that their full realization depends on a new paradigm that is consistent with a set of ideas emerging in scientific theory. This insight provided the last piece of the jigsaw that enabled me to decide on the nature of the book required to satisfy my agreement with Butterworth-Heinemann.

As a result the book describes a new paradigm and suggests the practical implications for construction. The decision to call the book *Construction – The Third Way* recognizes the complexity of today's world and the inadequacy of the old recipes for both managers and governments. The belief of the political right in free markets, individual freedom and competition, and the left's old preoccupation with state control, high taxation and producer interests, are equally flawed. The UK has experienced both extremes during the last 50 years and the results leave the construction industry, in common with the rest of the country, in need of a third way.

The third way, as described by Blair (1998) and Giddens (1998), is often denigrated by their political opponents as being no more than a wishy-washy compromise, lacking conviction and having no clear philosophy. In fact, the complexities of today's richly interconnected communities and their impact on the world's ecology lead Blair and Giddens to recognize the need for direct, inclusive democracy at all levels guided by free and open communication, and a focus on delivering better value for everyone who accepts the rights and responsibilities of belonging to a community. This takes them beyond the old political divides of right and left. Freedom and equality provide the classic dichotomy; both are desirable but they conflict. The freedom to pursue individual interests leads to inequalities, while attempts to impose equality limit individual freedom. Either freedom or equality pushed to extremes produces poor results. What is needed is a fair balance in the interests of the whole community. As Tony Blair says 'what matters is what works to give effect to our values'.

Construction – The Third Way accepts the importance of seeking consensus on values that are then given effect using the whole spectrum of human behaviours, from cooperation to competition. Applying these ideas to construction helps us to understand how interests that are traditionally seen as being in conflict can find win:win agreements. Producers, the consultants, contractors and specialists who design and construct new facilities can have fair profits, while at the same time consumers, the customers and communities who use the facilities can enjoy better quality and value.

The practical challenge is to combine Japanese efficiency based on steady, continuous improvements in performance with the ability widespread in the USA to start new businesses and create new jobs. But to do so in a way that avoids the weaknesses of both approaches. Japan's weaknesses are evident in its recent financial crises which reveal an inflexibility in continuing to invest in ever more productive capacity, even though the market has changed and customers now want different things. The weaknesses in the US approach are evidenced by its grossly unfair distribution of wealth, far too many children living in poverty in the richest nation on Earth, and too many people in low-paid, insecure jobs. Japan has relied too much on inward-looking cooperation and the USA relies too much on competition which reinforces inequalities. The third way balances cooperation and competition. It does this, not by wishy-washy compromise, but by basing decisions on the interests of the whole community guided by clear benchmarks of what is achievable. This requires team decisions, through communication that includes all people affected by the outcomes. The diversity of human situations and circumstances ensures that many different ideas will be tried and many different outcomes will result. The third way rejoices in diversity which comes from recognizing that communities at all levels are better off working with the richly interconnected networks that form human communities and their environments than by trying to impose the old certainties of free market competition or centralized management.

My conclusions about best practice for the construction industry reflect this third way. It is put into effect by teams making decisions about their work in ways that balance cooperation and competition. The teams cooperate with all organizations affected by the work, or they have targets and constraints that take account of these interests. A competitive drive for innovation and creativity comes from teams setting targets based on benchmarks of world class performance. On the evidence of the case studies I have undertaken over the past five years, teams that balance cooperation and competition in these ways deliver better value for customers and earn higher profits for construction firms. The term 'the third way' provides the most accurate description of this approach which seems to me to be the current best practice for the UK construction industry to adopt.

Writing this book has helped me to understand how these ideas all come together; my hope is that *Construction – The Third Way* will do the same for its readers.

Finally I am pleased to be able to acknowledge the many people who have contributed to the development of my ideas. These include many colleagues in the Department of Construction Management and Engineering, the staff of the Centre for Strategic Studies in Construction, many members of the Design Build Foundation and the Reading Construction Forum, including particularly its Partnering Task Force, and the editorial staff of Butterworth-Heinemann. I must especially thank my research assistant for the past five years, Sarah Peace, for her many valuable ideas and consistent help with my work. Most importantly, I am grateful for this opportunity to thank my wife, Sue, for her unfailing inspiration and support for my work.

John Bennett
The University of Reading

The new paradigm

1

Paradigm shifts

The word paradigm was originally defined by Kuhn (1962) as the views shared by a scientific community but Capra (1996) describes how it is now widely used to describe the concepts, values, perceptions and practices shared by any group of people. Thus a paradigm is learnt from experience of living and working in a community. A paradigm shapes the decisions we make and the actions we take. It determines how we see the world, other people and their behaviour.

Once people have learnt one paradigm they are reluctant to change it. They make such fundamental changes only under the pressure of major events. Thus a paradigm shift is a revolutionary break with an established way of viewing the world. Many parts of the construction industry face just this kind of pressure as rapidly changing technology leads to new demands from customers.

Information technology is changing the nature of most human activities. This means that the construction of different buildings and other facilities to accommodate the new kinds of behaviour is needed. The concept of an 'intelligent building' is already changing the way buildings are used; internal comfort conditions can be matched to the changing needs of users throughout the day. Similarly there is serious research into the feasibility of continuously monitoring urban environments to help traffic flows, policing and the emergency services, and care in the community.

The construction industry has to respond to these demands for new and more sophisticated products. However, more fundamentally, new technologies are consistent with different ways of manufacturing that rely on cooperative, long-term relationships. The new methods deliver significantly better value, faster and more reliably than traditional methods. This potential was first exploited in Japan because the new technologies fit their cooperative, group culture. The advantages first became evident internationally in the car industry where Japanese products justifiably gained a reputation for providing better quality and value. In response

Western car firms adopted the Japanese methods which are now commonplace in all main manufacturing industries. They already influence the leading edge of construction practice.

The main reason for this change is that most of the construction industry's major customers face pressures in their own businesses caused by these global changes. Customers have been forced to change the way they think about their businesses and the way they work. It is therefore not surprising that many of them expect the construction firms they employ to adopt similar new and more efficient methods. Recent research into leading practice makes it apparent that adopting such methods requires the construction industry to think differently about its work; that is, to make a paradigm shift.

Collapse of the management paradigm

It is significant that changes similar to those becoming evident in parts of the construction industry are already sweeping through many other industries in the West. This has happened because the assumptions on which Western managers have traditionally based their working methods produce inefficiencies wherever they are applied. The results have become increasingly unacceptable to customers. Slow deliveries, poor quality, high prices and broken promises are no longer tolerated. As a consequence, managers in every industry have been forced to make fundamental changes in the way they work. To do this they had to think about their work in a different way. In other words they made a paradigm shift.

The nature of the widespread change in management practice is described in a great mass of new books. For example, Locke (1996) describes the collapse of the American management mystique. He argues that the strengths of American-style management are no longer relevant. Its key features, analysing problems, giving instructions to subordinates and dealing with conflicts, provide an inadequate basis for dealing with today's world. Locke argues that hierarchical, top-down approaches need to be replaced by more inclusive, cooperative approaches of which Japanese-style management is the most widely quoted example.

Essentially the same case is argued by Lazonick (1991) who traces the development of management from its origins in the market-based, proprietary capitalism that emerged from the British industrial revolution in the eighteenth and nineteenth centuries. Its fundamental idea is encapsulated in Adam Smith's invisible hand. This is the belief that if everyone pursues their own interests, the market will ensure the best outcomes. The naiveté of this view became obvious as markets grew larger and technology became capital-intensive.

Market-based, proprietary capitalism gave way to managerial capitalism in a paradigm shift that occurred in the early years of the twentieth century. America invented management as a distinct responsibility and used its strengths to set up large hierarchical structures to plan and coordinate vertically integrated and mechanized production processes. Chandler (1977) chronicles this second industrial revolution which led to mass production guided by what he accurately calls the visible hand of managerial decision making. This has provided unprecedented riches for those lucky enough to live in developed countries. However, work designed and controlled by managers has become increasingly unattractive to affluent and well educated workers. It is also slow to respond to today's rapidly changing technologies and markets. The inevitable tensions limit what managerial capitalism can deliver and so the way was prepared for another revolution in production methods.

The new approach emerged first in Japan and, since the early 1980s, Japan has outperformed America in the production of consumer durables. This has been most noticeable in the production of motor cars, the twentieth century's most important industry, and electronic equipment, which is already of major significance in every aspect of our lives and is likely to be the crucial industry of the twenty-first century.

Japan's success is based on what Lazonick (1991) called 'collective capitalism'. This relies on cooperation between tightly knit groups of firms. Decisions are made by consensus in networks that spread throughout multi-firm organizations and which include customers and suppliers. Government, too, is often deeply involved in the decision making of these cooperative networks.

Even more recently, the leading edge of international trade has become dominated by computer-based service industries centred on the USA which, as a result, has enjoyed a remarkable period of economic growth in the second half of the 1990s. Software and its many applications has become more significant in creating new businesses than has manufacturing. Construction, although it needs to add sophisticated services to its products, remains primarily a manufacturing industry. Hence the important lessons for the UK construction industry come more from Japan's strengths in manufacturing than the USA's strengths in software and service industries. Amongst the major developments in manufacturing, the identification of lean production is particularly important in understanding what needs to happen in the construction industry.

Lean production

Womack *et al.* (1990) describe in fine detail the move from traditional American management to Japanese cooperation at the leading edge of the motor car industry. They call the new approach 'lean production' because of the central importance placed on identifying and eliminating waste. They define waste as processes that add no value for customers. Japanese firms use lean production in their highly efficient production of motor cars that customers find increasingly attractive. Womack *et al.* explain how first Toyota, and then the other major Japanese car producers, brought their workforce and then their customers and suppliers into their decision making processes. In so doing so they abandoned the American model of hierarchical management in favour of an entirely new approach that relies on building cooperative long-term relationships. The resulting lean production now dominates car production throughout the world.

Fundamental changes of this magnitude take place when conditions are right for the new approach. Given the right conditions, the change is self-reinforcing as the new approach sustains the factors that allowed it to emerge in the first place. This is now evident as the emergence of global markets and rapid developments in technology, especially information technology, are first causing and then reinforcing fundamental changes in management practice. The developments allow major firms to search for the lowest-cost reliable suppliers, wherever they happen to be based, so that much basic work has been moved offshore away from developed countries. In these same firms, layers of middle managers formerly employed in routine information processing have been made redundant by information technology. So those that remain are engaged in far more communication, most of which takes place through the Internet, mobile phones, faxes, video conferences and other devices handling digital data. Head offices are much smaller and many firms own little except information and networks of contacts.

These ways of working miss the unspoken but important messages that come from body language and other aspects of old fashioned face-to-face communication. So managers need new skills in building more secure relationships that can function at a distance. Therefore, ideas about cooperation and trust have become widely discussed. Managers have to recognize when work needs face-to-face relationships; hence the wide use of workshops, project offices and other arrangements that bring teams together to tackle specific, difficult problems.

This coincidence of multi-faceted, self-reinforcing changes which affect many aspects of a community and its work is evidence of what is properly called a paradigm shift. The specific change described in this book is reinforced by fundamental changes in the way science views the world. This scientific revolution has emerged over the past century but its key ideas have broken through into popular literature only very recently. These scientific ideas call into question many of the assumptions that underpin the management paradigm. Also the ideas suggest new patterns of working that are more in tune with human nature and the world we inhabit than any traditional management-based approaches.

A new view of the world

Over recent decades science has built a picture of the world and our relationships with it which is very different from that which provides the intellectual basis for the methods and institutions used by managers in Western, market economies. As these new ideas are understood and discussed outside of the scientific communities, they are giving rise to new theories about management. Some of the theories anticipate and reinforce changes already taking place in practice.

The key ideas of the new view are that our world consists of richly interconnected networks in which ideas of hierarchy are human projections not justified by the structures and processes found in nature. Competition is not the main driving force of change and evolution; it is the exception and generally provides an unsustainable basis for any species. Cooperation is much more widespread and important in explaining the evolution of life on Earth. As Dawkins (1986) puts it, life depends not on the survival of the fittest but of the 'fittingest', by which he means that the species that have survived are the ones able to cooperate best with their environments. Capra (1996) uses this new view of the world in explaining how cooperation and symbiosis have been central to the evolution of life on Earth.

Science now sees the world, including all living creatures, as one incredibly complex system of networks in which feedback loops give the whole and individual parts the power of self-organization. The properties of every part of this vast web are determined by other parts with which they interact. *Figure 1.1* illustrates in a greatly simplified form the general nature of this view of the world.

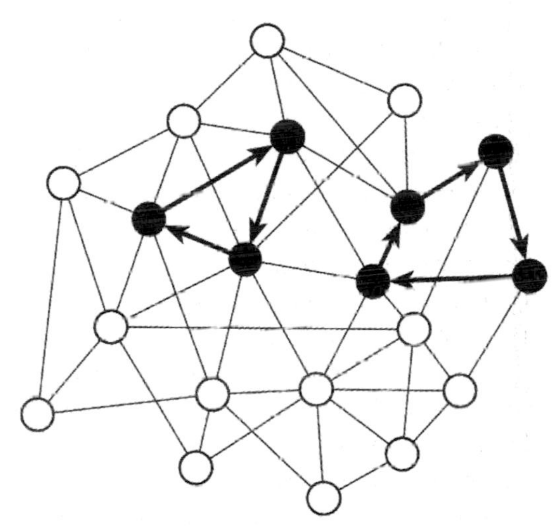

Figure 1.1: Network with feedback loops

Interdependence influences all the interactions between humans and the environments in which they live which, of course, includes other human beings. What this means is that the world we experience is determined by what we choose to regard as distinct things or events and our perceptions of them. We experience not nature, but nature as defined by our method of perception. Other species experience a different world because their perceptions are different. They literally see things we cannot see, hear things we cannot hear, smell things we do not know exist, and so on. Equally we see, hear, touch, smell and taste things that other species are unaware of.

In a similar way other human cultures experience a different world from ours. These differences are reflected most completely in the languages used to communicate. This is true for cultures based on race, religion, nationality, age, social organization, company or construction project. Each of these cultures selects parts of the richly interconnected world to give attention to, regard as important or ignore.

Having decided what should be regarded as separate and distinct, people develop ideas about the relationships between these things and events. The resulting theories guide their decisions.

Any given world view is centrally important in defining the nature of the people who share it. Which is why great efforts are made to defend any given culture. Wars are fought because people choose to see the world differently. Legal battles rage because of clashes of culture which give rise to different views about cause and effect. Children are taught, mainly by the example of adults, a great variety of different views about right and wrong. Managers in different firms working on the same project often see any given sequence of events very differently and their different perceptions create conflicts over rewards and blame.

In the construction industry, disputes are common and their incidence and the particular form they take result from current theories about management hierarchies, market competition, contractual rights and responsibilities, and much else that guides the thinking of managers in the West. The view that lies behind these theories conflicts with the picture that science now provides of humans and the world we inhabit. These errors of perception lead mangers to see a world growing ever more complex and uncertain as day-to-day experience contradicts traditional theories about the way the world should behave.

Complexity and uncertainty

As a result of the mismatch between theory and experience, the choices facing customers and managers involved in construction appear bewildering. The apparent complexity and uncertainty are reflected in practical actions. For example, there are many different procurement routes currently in use in the UK construction industry, each strongly supported by practitioners who have built successful businesses by learning how to cope with the conflicts inherent in one particular approach. Thus, there are a multitude of independent specialist designers as well as a variety of multi-discipline design studios offering many different services to customers. There are general contractors, design build contractors and management contractors. There are construction managers who offer their services on a consultancy basis in competition with these various kinds of main contractor. In addition, there are specialist contractors who offer widely differing combinations of design, manufacture, construction, commissioning and maintenance services in respect of a bewildering variety of construction technologies. Then there are consultants, including project managers, quantity surveyors and facilities managers, all of whom offer to help customers deal with the choices generated by all the other specialists.

As a result there is a massive literature describing the many different procurement routes. Much research is devoted to defining and measuring the performance of the various alternatives. As a result, the strengths and weaknesses of different approaches can be described in theoretical terms. Best practice is identified and codified on the basis of those project case studies judged to be successful because they encountered fewer problems or delivered better results than the norm. However, applying the resulting ideas in practice has become virtually impossible because the underlying paradigm causes any one construction project to appear more complex and uncertain than the theories and best practice guides assume it should be.

Elements and hierarchies

The view which gives rise to these concepts of complexity and uncertainty sees the world as made up of independent elements that form hierarchies.

In this view, construction activities are seen as independent in the sense that each of them can be carried out in isolation by specialists. The specialists' work is arranged sequentially. Indeed, in traditional craft work the separate crafts do not need to meet. Each craft undertakes its work independently and leaves the finished work in the form expected by the next craft in the sequence. So bricklayers leave window openings in the form expected by carpenters who provide the fixing grounds for joiners, who are followed by glaziers and then painters in a sequence determined as an integral part of craft training. *Figure 1.2* illustrates the model implicit in this view of construction as a sequential network.

In a similar way the work of the traditional professions is designed to allow each discipline to work independently. This independence is seen as a strength as

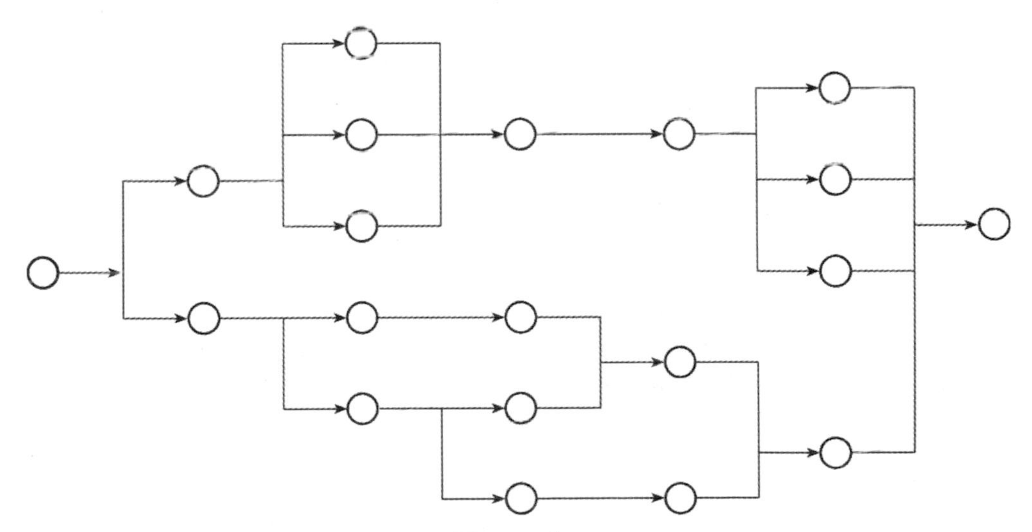

Figure 1.2: Construction seen as a sequential network

professionals argue that they can provide customers with independent advice. The fact that each element of the resulting advice is partial and in total includes many contradictions is a problem only for the customer, not for his or her independent professional consultants. The construction professions are well practised in the complicated games that result from their separate and independent agreements with their customers (or clients, as they prefer to call them).

The organizational arrangements implicit in the old paradigm are riddled with perceptions of hierarchy of the kind illustrated in *Figure 1.3*. Each independent activity has its own hierarchy and only the top levels are supposed to communicate with each other. So, each specialist contractor involved in a project is required to nominate one manager to be responsible for his firm's work. Communication is then channelled through the named managers. Direct contact between other levels is seen as bad practice and is forbidden in many standard forms of construction contract.

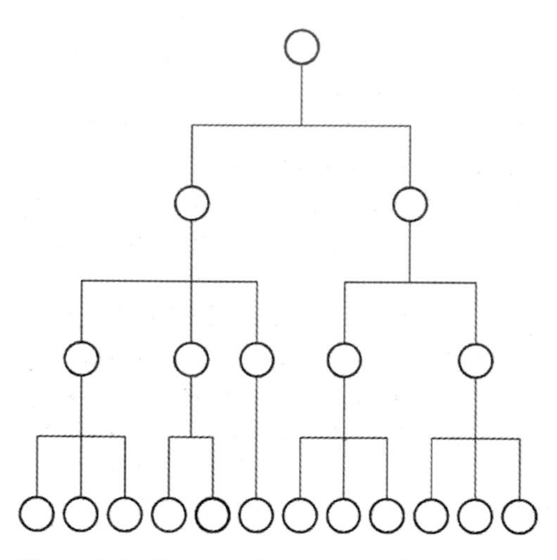

Figure 1.3: Construction seen as a heirarchy

The concept of hierarchy is more pervasive than this and indeed dominates thinking about relationships throughout construction projects. In building projects in the UK, for example, there is a widely recognized hierarchy of disciplines. Architects sit proudly at the top and craftsmen and their supporting labourers struggle at the bottom. People know where they fit in the hierarchy and defer to those above them. Thus, in traditional practice, architects' judgements and decisions are rarely questioned by other professions. Yet it is accepted as reasonable that architects should make judgements about the timeliness and quality of other peoples' work. If an architect's judgement about another's work is challenged it is seen as a very serious problem, and most forms of contract include elaborate procedures to deal with what is classified as a formal dispute. Most formal disputes in construction arise in this way and are based on conflicts resulting from the different perceptions of the independent professions.

Conflicts have become a serious problem in the UK construction industry and report after report complains about the incidence of disputes. Latham (1994) provides a well argued and influential report that makes detailed proposals for ending the adversarial attitudes engendered by disputes. Its proposals, however, are framed within the old paradigm and so, although it makes sound recommendations about teamwork and partnering, it sees the way forward largely in terms of better contracts, more clearly defined responsibilities, a broader evaluation of competitive tenders and similar ideas grounded in the management paradigm. More fundamental changes in the construction industry are required before its customers will get the standards of quality, speed and efficiency they expect from other businesses.

Change in construction

The ideas described in this book are based on the view that the problems described by Latham (1994) and others are caused not by the nature of construction but by the way managers in the industry choose to view their work. So to move forward from the perceived complexity and uncertainty caused by a paradigm based on elements and hierarchies requires managers to change the way in which they see construction. It is not customers, nor designers, nor specialist contractors who cause the problems. The construction industry's poor performance derives from managers throughout the industry taking an old-fashioned view of their work and, in so doing, creating frameworks that force customers and technical specialists to waste time and resources on unproductive activities.

This is the situation that has faced managers in most industries during the past two decades. Some, including some in construction, have made the necessary changes. The results seen in other industries have been described by many authors. For example, Oliver and Wilkinson (1992) and Womack and Jones (1996) leave no doubt about the benefits of making the paradigm shift nor about the deeply ingrained resistance to change on the part of those still attached to the old ideas.

The argument on which this book is based is that the world appears to be complex and uncertain to managers of construction projects because they choose to view it in terms of elements and hierarchies. The book uses a wide range of recent research into leading practice in construction to propose a more useful way of seeing the world so that what is all too often seen as complex and uncertain can be recognized as manageable.

Practical theories

Two sets of ideas are particularly useful in understanding the practical implications of the new paradigm. They are systems thinking and chaos theory. Together they provide a framework of ideas that are implicit in the long-term strategies of really successful organizations. These ideas help managers make good decisions even though they cannot predict all the problems they will face, nor the source of good answers.

A useful introduction to the practical implications of these ideas can be found in De Geus (1995) which was written to help the Shell Group plan its future strategies. De Geus looked at the histories of companies that had existed as large enterprises at the end of the nineteenth century. An intensive search identified only about forty that still exist today with their corporate identity intact. All the rest had been swept away by changing markets or technologies that they had been unable to deal with.

The strength of De Geus' work is in his finding that the surviving companies share key characteristics which enable them to be efficient and yet cope with major changes. *Figure 1.4* illustrates the most important characteristics of the surviving companies. They sustain and steadily develop their physical and human resources in their mainstream businesses and, at the same time, tolerate marginal activities. These are usually pursued by small dedicated groups of enthusiasts who believe they have spotted a potential opportunity or simply want to develop an interesting idea. It is the existence of activities at the margin that allow companies to change direction as markets and technologies change. In contrast, companies unwilling to tolerate activities that consume resources outside of their mainstream business are often destroyed by the stress of making fundamental changes under pressure from market forces or technological innovations.

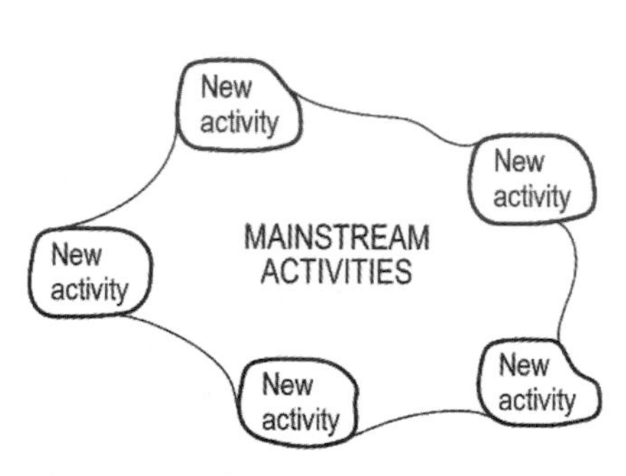

Figure 1.4: Companies that survive long-term

In the short-term, tightly managed companies are often more efficient at their mainstream business than their more tolerant competitors. However, centralized companies are not flexible and De Geus' study shows that during the twentieth century, centrally directed change has been difficult to manage. Companies tolerant of activities at the margin of their businesses have a better track record in responding to change.

Tolerance is needed because there is no way of predicting which marginal activities will provide the best response to some future threat. All that senior management can do is ensure that they recruit talented people and then give them the tools and opportunities to understand specific parts of the company's environment. Most will concentrate on the mainstream business but some will get excited by other ideas and they must be allowed to pursue them inside the company. In the short term this is wasteful of resources but, in the long term, De Geus' study tells us that tolerance for marginal activities gives organizations the flexibility needed for long-term survival.

Thus managers need to reconcile the demands of short-term efficiency and long-term flexibility. Systems thinking and chaos theory provide ideas that help achieve this difficult balance.

Systems thinking

Systems thinking is at the heart of the paradigm shift that is radically changing the theory and practice of management. Systems thinking is one of the great scientific advances. It gives us a way of thinking about complex things as a whole. This is in direct contrast to the established scientific paradigm which divides complex things into smaller and smaller parts in the belief that the way to understand the whole is to understand each of the elements.

A common definition of systems, which for example runs through Open Systems Group (1981), is that they are any group of elements which relate to each other in a sufficiently regular manner to justify attention. This rightly emphasizes the need to choose a specific viewpoint in using systems thinking. This is based on the belief that the way to comprehend complex things is to concentrate on one characteristic. For example, an organization can be regarded as a decision making system and attention can be concentrated on actions and interactions that contribute to decisions. Every feature not involved in decisions is ignored. This very focused view enables the complexity of a whole organization to be understood. So, having understood how decisions are made the same organization can be viewed as a system of formal authority, a system of information flows, a system for processing raw materials, or from many other systematic viewpoints. Each enables the whole organization to be understood but from one specific point of view.

Capra (1996) adds precision to the concept of systems thinking in defining a system as an integrated whole whose essential properties arise from the relationships between its parts. He explains that systems have organized complexity that can be viewed at distinct levels which have properties that do not exist at other levels. So the essential properties of a system are properties of the whole which are not possessed by any of the individual parts. To take a simple example, sugar is made up of carbon, oxygen and hydrogen atoms, none of which taste sweet. Sweetness is an emergent property of the system of atoms we call sugar.

Taking the view that systems cannot be understood by discovering the functions of the parts is a dramatic break with ideas that have dominated Western thinking for at least the last 200 years. It is now becoming widely understood that what matters is the organization of the parts, that is the pattern of relationships between them. The behaviour of parts can be fully understood only by considering them in the context of a larger whole. This is the essence of systems thinking which focuses attention on the pattern of things rather than on their physical structure. Elements cannot be understood in isolation, their context determines their behaviour. As a result, systems thinking focuses attention on interactions between parts and their environments. This applies equally to whole systems which can be understood only by relating them to the larger whole which forms their environment. Thus the sweetness of sugar does not exist until it interacts with our taste buds to produce what we regard as a sweet sensation.

Systems thinking undermines the traditional scientific paradigm which seeks to understand the behaviour of complex systems by studying the properties of its parts in isolation. This is important because the analytical approach provides the justification for many influential ideas in economics, politics and management. As indeed it does for many of our social institutions. It is gradually being recognized in all these fields that the powerful new insights provided by systems thinking call established ideas into question.

The key change is understanding that the world is made up of networks that are richly interconnected. The connections form feedback loops which turn the networks into systems with the capacity to respond and adjust to change in ways that give rise to new structures and forms of behaviour which ensure the survival of the system. Capra (1996) draws on a wide range of new scientific ideas that have emerged from systems thinking to show that the world as whole has this capacity to self-organize; as do many systems within the world, including human organizations. Accepting this view means we should see organizations as social systems bound together by communication in which feedback working within control loops plays a key role. *Figure 1.5* indicates the richness of feedback loops needed in even a very simple network.

Seeing an organization in this way means we need to understand its pattern, that is the essential configuration of relationships between the elements. The pattern of an organization is distinct from its physical reality. However, a comprehensive understanding of an organization deals with form and substance, that is with pattern and physical form. The physical form has a structure which is sustained by processes. In effective organizations, both structure and processes are guided by the ideal pattern. Therefore a comprehensive understanding of human organizations needs to include all three characteristics of living systems: pattern, structure and processes.

Viewing organizations in this way forces us to work with their capacity to self-organize. The world is too richly interconnected to avoid unexpected and probably damaging outcomes by using any other approach. Therefore management should proceed, not through exercising control and domination which is at the heart of established management theory, but on the basis of respect, cooperation and communication.

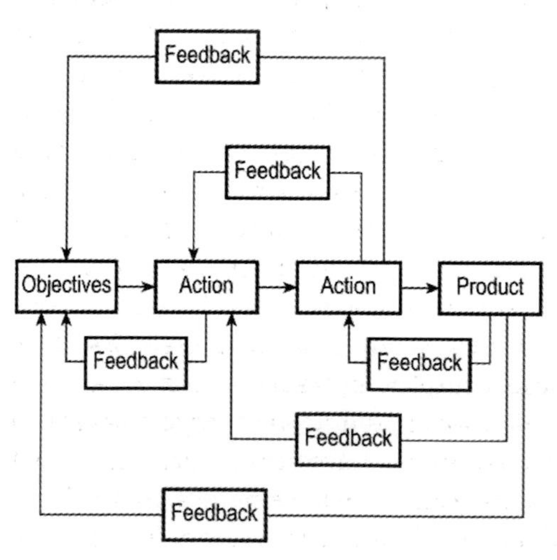

Figure 1.5: Feedback loops

It is useful to think in terms of systems because they share characteristics which scientists can describe in precise terms. So, faced with a new system, a scientist or a manager can draw on carefully researched descriptions of these common characteristics to help understand its behaviour. This means that much systems thinking consists of deciding which other well understood systems a new system is most like. Thus, traffic engineers have gained new insights into the behaviour of traffic on motorways by using ideas generated in studies of the flow of water in river systems. Another important example is provided by Beer (1972) which identifies the key characteristics of the human nervous system when it is viewed as an information processing and decision making system. This provides new insights for managers into how information is processed and decisions made in human organizations.

The idea of using one system as a metaphor for a superficially different one is explored in great depth by Morgan (1986). In practical terms metaphors provide managers with a range of different ways of thinking, any of which may prove to be useful in solving problems.

Morgan uses metaphors to review the main bodies of established management theory. This shows that established management ideas are breaking into diverse theories and different metaphors are implicit in the various schools of thought. Morgan characterizes each theory in terms of: machine, organism, brain, culture, politics, psychic prison, flux and domination. He suggests that each of these metaphors may suggest useful models and ways of thinking for managers as they tackle difficult problems. Morgan acknowledges that other metaphors may be equally useful and takes the view that it is for managers to develop their own ways of thinking so they become skilled at using a repertoire of metaphors to guide their work.

Chaos theory

We need one more important idea before the scene is set for a full exploration of the implications of the new paradigm on the work of managers responsible for construction. This is chaos theory which emerged from the work of scientists attempting to understand the behaviour of large, complex systems. Some of the key ideas came from The European Centre for Medium-Range Weather Forecasts based in Reading, UK. By chance scientists found that small changes in their input data describing current weather conditions normally produced small effects on the resulting medium-range forecasts. Occasionally, however, the results of such small changes in their input produced massive, chaotic differences in the forecasts.

Similar effects have been found in many other situations, including in various stock markets and the market for components in the world's car industries. Scientists have constructed beautiful models of these complex systems which reveal the underlying patterns in their behaviour. These models tell us that large systems in which the elements are richly interconnected have mainstream processes that behave reasonably predictably. However, occasionally these same systems behave chaotically. It is possible to predict the likelihood of chaotic behaviour but its precise time and form are unpredictable.

Thus medium-range weather forecasts can be made with a reasonable degree of confidence most of the time. However, occasionally the forecasts vary so widely that no forecast is possible. The world's weather systems have entered a period of potential chaos and the best meteorologists armed with the world's biggest computers cannot predict what will happen. Human organizations that are large systems in which the elements are richly interconnected have these same characteristics. Most of the time they behave predictably but at times chaos will reign.

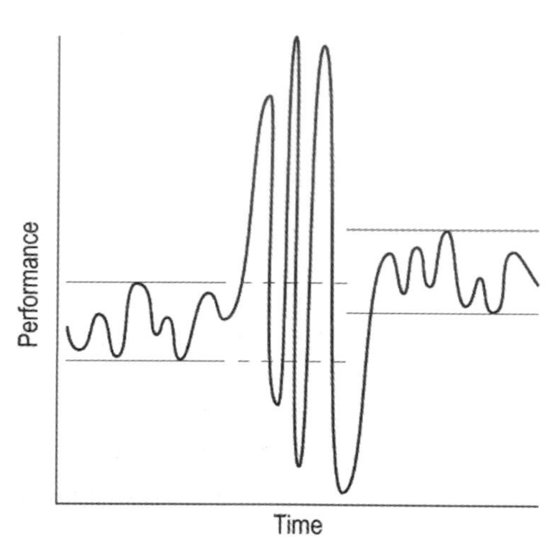

Figure 1.6: Control and chaos

Many human organizations, including some closely involved in the construction industry, are of precisely this nature and managers need to be aware of the possibility that the performance of their organizations may behave like that shown in *Figure 1.6*. Following a period in which the organization's behaviour is controlled, it becomes wildly erratic for a time, only to return to control, often at a different level of performance. Hence managers should expect construction to behave predictably in its mainstream activities but throw up surprises at the margin. Managers have no choice but to plan and act within these dual behaviour patterns. It is sensible to aim at controlled efficiency in mainstream work but flexibility is needed to cope with unpredictable change.

Control and flexibility in construction

At this point it is helpful to consider the way in which control and flexibility are managed in the world's leading construction industries. Atkins *et al.* (1994) provide a useful basis for this. Their work describes a major study for the European Union which was designed to provide the basis for an EU policy on construction. Amongst much else it provides carefully researched descriptions of the construction industries in Europe, the USA and Japan.

Europe

In general, construction industries in Europe have paid too much attention to the need for flexibility. Small specialized firms provide an almost infinitely flexible industry in which project teams can be assembled, disbanded and reformed at the whim of customers, consultants and contractors. All of these key actors are guilty of failing to see the need for mainstream stability which provides the basis for efficiency. They have allowed marginal unpredictability to dominate their thinking and have created structures and processes designed for one-off creativity. The resulting high costs are largely accepted as inevitable.

Market forces provide the most effective way of organizing such an industry. The small specialist firms assembled to undertake individual projects are best organized by fitting them into predetermined hierarchical structures and requiring them to work to implicit procedures learnt in their craft and professional training. This is how most of the world's construction industries are organized but the wildly variable performance that results is increasingly unacceptable to customers.

USA

To satisfy the need for control the US construction industry has superimposed standards on a market-based approach but the search for efficiency has, to a large extent, served to drive quality standards down.

The most characteristic approach of the US construction industry, construction management, displays its key strengths well. These are standardized buildings produced quickly and efficiently. Key factors in providing this performance are architects' and engineers' standardized design details and construction managers' cost and time control systems that allow specialist contractors considerable freedom to choose their own preferred construction methods. Specialist contractors are selected on the basis of fierce market competition and, once appointed, work within the designer's overall design concepts and the construction manager's programmes and budgets. These project-specific controls make extensive use of local procedures that are well understood by the specialist contractors operating in that market. Standardized materials and components are readily available at short notice. The specialist contractors are self-sufficient in the sense of providing all the supporting plant and equipment they need to carry out their work efficiently. They take responsibility for negotiating the interactions between their work and that of other specialists. Much of this takes place on site in deals agreed day-to-day by the specialist contractors' foremen.

Provided that everyone involved sticks to this established way of working, customers get their new facilities quickly and at low costs. Generally when the facility is handed over to the customer there are too many defects and overall quality is poor. The approach is not flexible. Once customers or designers move outside the limits set by the various standards, the industry's performance is weak. Projects using creative, individual designs tend to finish late, give rise to many disputes and claims for extra money, and litigation is a major sore disfiguring the industry's relationships with its customers.

Japan

The Japanese approach delivers good quality efficiently by using a very standardized approach. This approach is based on processes developed over many years by large design build firms. The processes are applied with great consistency on individual projects but include procedures aimed at continuously improving all aspects of the firms' performance.

The efficiency of the processes depends on long-term relationships between customers, design build contractors, suppliers, specialist subcontractors and sub-subcontractors. These relationships operate within tightly knit families of firms that have existed for decades and work on the basis of cooperation and trust.

There is competition between the families of firms within a carefully structured market. Established relationships with long-standing customers are rarely challenged. The Government allocates public sector work on the basis of an elaborate market ranking of firms which is up-dated and published annually, but which tends to reinforce the established order. Also there are many joint ventures and widespread collaboration on research between contractors. The result is that the market share of individual firms remains remarkably stable year after year, not least because it is considered poor behaviour in Japan to disturb the market. Work for a new customer or for one who builds occasionally is the subject of fierce competition. Firms will bid low and provide an excellent service in the hope of establishing a new, long-term, loyal customer. However, new customers are the exception and well established, long-term relationships based on cooperation and trust shape the Japanese construction industry.

Design build contractors work closely with regular customers to identify opportunities where new construction will help their businesses. They also work closely with and take responsibility for the well-being of subcontractors. They set tough standards and invest in helping the subcontractors achieve them. They set prices which ensure that subcontractors earn a fair profit and so can afford to concentrate all their efforts on doing the best possible work. Subcontractors strive to complete the agreed work on time exactly to the specified standards. They are willing to try new technologies and actively look for ways of improving quality, safety and productivity. They use these same cooperative methods in working with their sub-subcontractors.

The most important actors in the Japanese building industry are the major design build contractors. Pre-eminent among these are the big five, Kajima, Ohbayashi, Shimizu, Taisei and Takenaka. They play a leading role in establishing the methods of the industry and representing its interests to Government. In return major, international-sized Government projects are shared between the big five, usually on a joint-venture basis.

The big five design build contractors are large, each employing in excess of 10 000 engineers. They provide comprehensive design and construction management services. They routinely find land, arrange finance, produce conceptual and detail designs, manage the manufacturing and construction processes, and repair and maintain the buildings they produce. Indeed, they will do whatever a customer needs, within the predetermined budget and programme so as not to risk losing an established customer.

There are many other construction firms of various sizes from very large to small but the Japanese approach is most clearly evident in the work of the big five working on building projects for their regular customers. These customers are well informed because they routinely share information about construction performance, costs and times with other firms working in their market sector. This means they can define their construction requirements in detail, including specifying the budgets and programmes they expect their contractor to achieve.

The big five have consistent processes, based on tried and tested methods of working, which are used on all the firm's projects. Staff are encouraged to search for ways of improving these established methods by meeting regularly in quality circles. In addition, new ideas and potential major improvements are researched and developed in large and very competent in-house research institutes. When a new and better method is identified, either as a result of ideas developed within a quality circle or as a result of research and development, it is discussed widely with all those likely to be affected. Only when there is a wide understanding of the change is it introduced. Then it becomes the firm's standard approach until a further new and fully considered better answer is found. By providing for change in this way the Japanese design build contractors combine steady, reliable efficiency with continuous improvements in their established approach.

The resulting methods are described in more detail in Chapter 6. The key point at this stage is that the overall pattern of long-term, cooperative relationships throughout the industry, with customers and Government, is consistent with the new paradigm. In other words managers in the Japanese construction industry act in ways that are consistent with the picture of construction shown in *Figure 1.7*.

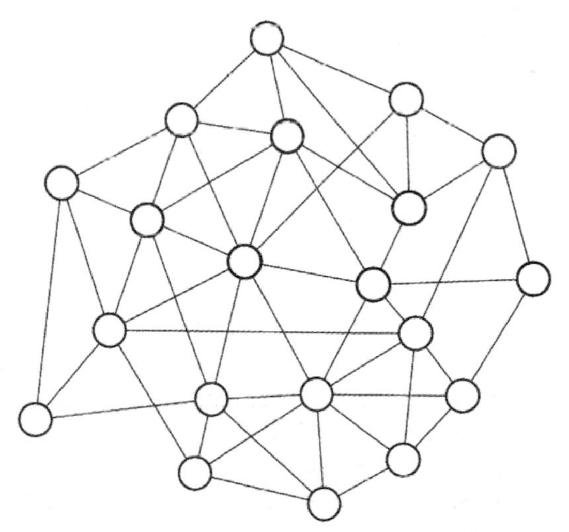

Figure 1.7: Construction seen as networks of networks

However, in the later half of the 1990s Japan experienced severe financial problems caused by over-investing in productive capacity for which there was insufficient demand. In no small part this was because other countries responded to the challenge from Japan and improved their own productivity (the car industry being one example). These events have caused major problems for Japan's construction industry which is geared up for highly productive work in stable markets. This does not invalidate the Japanese management approach, which continues to provide the basis for most manufacturing industries around the world. What, however, is called into question (as well as Japan's financial management) is the degree of flexibility provided by their management approach. It is outstandingly good at continuously improving an established approach but seems ill-prepared to deal with major change.

Comparative efficiency

The various national approaches result in very different levels of efficiency. Evidence from other industries suggests that Japan's cooperative approach is in tune with modern technology and, as a result, is efficient. If the same is true for construction we should find that international comparisons will show Japan to have the lowest construction costs.

Research into comparative cost levels as part of the strategic study for the European Union undertaken by Atkins *et al.* (1994), showed exactly that. The comparisons are based on pricing consistent samples of buildings and civil facilities in each country and adjusting for purchasing power parity to establish the comparative efficiency of the various national construction industries. *Table 1.1* is based on this research but takes account of other studies which reinforce and help explain the results in order to provide a fair picture of the comparative efficiency.

For many people the most surprising feature of the results is that they contradict the conventional wisdom that construction in the USA is efficient. A study carried out for the British Airport Authority throws useful light on why practical experience believes construction in the USA is cheap but an objective study shows it to be inefficient. The British Airport Authority's study, reported in Lynton (1993), took a building designed and constructed in the UK and had it priced in the USA. Care was

Table 1.1: International comparisons of building costs at purchasing power parity (UK costs = 100)

USA	Individual design	Normal quality	110
UK	Individual design	Normal quality	100
Europe	Individual design	Normal quality	95
Japan	Individual design	Normal quality	75
USA	Standardized design	Lower quality	75

Sources: Bennett *et al.* (1987), Hordyk and Bennett (1989), Lynton (1993), Atkins *et al.* (1994) and Vermande and Van Mulligen (1999).

taken to select a region in the USA where input costs were very similar to those in the UK. Thus the pricing largely reflected differences in efficiency. The building's price in the USA was almost 10 per cent higher than in the UK, which confirmed the European Union data about relative levels of efficiency.

However, the team that priced the building in the USA told the British Airport Authority that they would not have designed the same building. Normal US practice would have been to use many more standard components and to accept lower safety and comfort levels. So the building was redesigned to normal US standards and the price was about 75 per cent of average prices in the UK. Hence customers in the USA get standardized designs and lower quality at a lower price while UK customers get individual designs and higher quality at a price which is lower than that for an equivalent building in the USA but higher than that for a standard US building. Customers in the UK are not offered the option of cheap, standardized facilities. Customers in the USA are not offered the option of good quality individual designs at reasonable prices.

The Japanese results from the European Union study are supported in detail by the findings of a study visit to Tokyo by a group of very experienced UK practitioners. As part of the study visit each member of the group collected productivity data on Japanese practice in their own area of expertise. This was done at first hand so the experts could see the reasons for differences between Japanese and UK performance. The UK practitioners visited offices, factories and construction sites. In all these situations Japanese productivity was typically double that for equivalent work in the UK. As reported in Hordyk and Bennett (1989), this was not due to more sophisticated machines or equipment but was due to Japanese methods being highly refined by continuous improvement over many years. These differences in labour productivity are very similar to those reported in Bennett *et al.* (1987) and fully account for the European Union study finding that Japanese purchasing power parity costs are 25 per cent lower than equivalent UK costs.

Having said all of this, there are considerable problems with international comparisons of construction costs and the results are often controversial. A good description of the problems is given in Vermande and Van Mulligen (1999) which raises important issues and provides some new data. These have been taken into account in producing the values in *Table 1.1* which gives a picture of relative efficiency from a range of sources that broadly are mutually consistent.

The values tell us that Japan has the most efficient construction industry. Its cost levels are matched only by the US construction industry in its production of highly standardized, relatively low quality buildings. Comparing like with like the construction industry in the USA is marginally less efficient than those in Europe and substantially less efficient than that in Japan. These results support the idea that the benefits found in other industries from adopting more cooperative methods, apply also to construction.

The costs and benefits of interactions

The fundamental reasons for these differences in efficiency can be illustrated by reference to *Figures 1.2, 1.3* and *1.7*. Each shows twenty elements but very different numbers of interactions. The hierarchy shown in *Figure 1.3* needs only nineteen links to describe the interactions which managers wish to allow between the twenty elements. The sequential network in *Figure 1.2* recognizes twenty-six interactions; however, the network in *Figure 1.7* includes fifty links to describe the interactions between the twenty elements. This number of links is an arbitrary choice from amongst the 190 potential links in a network of twenty elements. Not all of the 190 links will be useful in any specific situation. *Figure 1.7* may provide a realistic picture of practice at any one point in time; at other times, other links will become important and so the picture would change.

The point is that the sequential networks and hierarchies of the old paradigm are designed to restrict where people look for information about their work, where they seek answers to problems and where they choose to supply information about what they are doing. The restrictions on communication result in disputes and misunderstandings. People waste time and resources on unproductive work, often because they are not told that others have moved on to a different plan. Even more seriously the traditional, restricted view of which interactions are legitimate cuts people off from the feedback that is essential for learning and improvement. The fundamental cause of these weaknesses is that the old paradigm causes managers to see interactions as costs and as a sources of potential threats. It therefore makes sense in their view of the world to limit and restrict interactions.

The new paradigm causes managers to see interactions as sources of help and information. It accepts that managers cannot predict where teams will need to look for good answers and information about their work. Therefore the manager's best course of action is to let well motivated teams interact with whoever they choose and to do everything possible to make the interactions as efficient and effective as possible. In particular managers need to ensure that feedback loops develop to inform everyone about how well targets are being met.

Figure 1.8 illustrates why the new paradigm leads to lower costs. It assumes that there is a given cost for undertaking basic work. In other words well motivated teams provided with all the information, materials, components, plant and equipment needed to do basic work will produce broadly the same costs. Therefore variations in total costs come from the costs associated with interactions and, of course, from the costs associated with other, not directly productive, activities.

The costs include the time taken up in interactions that could otherwise have been used for directly productive work. They include the cost of resources needed to facilitate interactions, including information technology links, meeting rooms

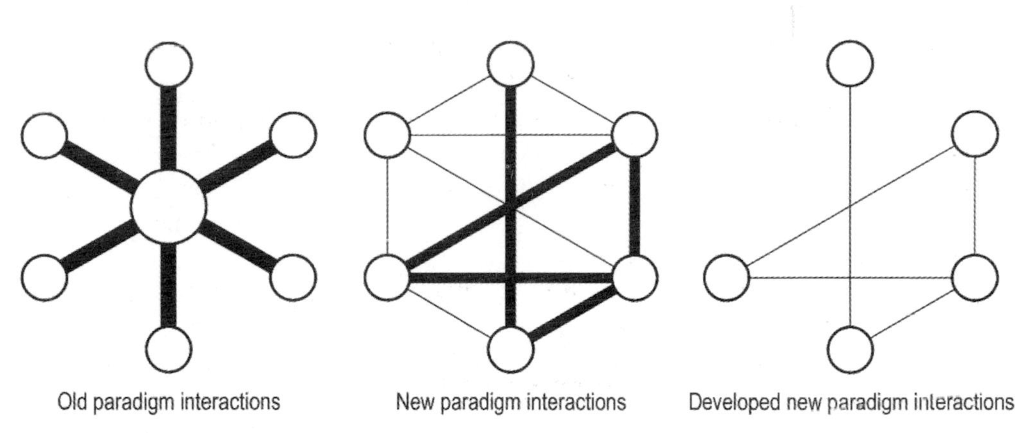

Figure 1.8: How interactions influence costs

and so on. Also costs arise from defining and restricting the interactions allowed to basic work teams, for example, defining contract conditions, giving formal notices, ensuring that information is complete before it is communicated and similar actions that are not directly productive.

Figure 1.8 shows how the new paradigm leads to lower costs. It begins with the old paradigm approach in which interactions are controlled and channelled through pre-defined routes. Then it shows the new paradigm allowing many more interactions and eliminating the centralized management of interactions. The interactions that teams find useful grow strong through frequent use and eventually are all that are needed.

In other words, the new paradigm reduces costs by allowing teams to decide for themselves where and when they need to interact with others. Well motivated teams will naturally keep interactions to the minimum needed to do their work. They will tend to use simple and direct forms of interaction which generally cost less than formal methods. Empowering the basic work teams also avoids the costs involved in predetermining, defining and restricting their interactions. When teams continue to interact over several projects and over time, the costs reduce still further as they learn to deal with each other at a more instinctive level and so need fewer and shorter interactions. In this more mature approach they also identify activities that add nothing of value to their work and eliminate them.

This pattern of costs is further illustrated by expressing the three diagrams of *Figure 1.8* in terms of hypothetical costs:

Old paradigm			
Basic work	6 × 10	60	
Coordination		20	
Links	6 × 10	60	140

New paradigm			
Basic work	6 × 10	60	
Main links	5 × 5	25	
Minor links	6 × 3	18	103
Developed approach			
Basic work	5 × 10	50	
Developed links	5 × 4	20	70

It is entirely significant that this pattern of hypothetical costs reflects the research results shown in *Table 1.2* which are described in the next section.

Changing to the new paradigm

The problem faced by traditional managers in accepting the new paradigm is that they have to give up control and empower teams to make their own decisions. This is difficult because the old paradigm taught them not to trust basic work teams.

The old paradigm, which stands in the way of necessary changes in construction, sees the proper form of organizations as a hierarchy of senior managers, middle managers, first line managers and basic workers. It regards individuals as responsible only for themselves. It sees economic growth as the natural and desirable result of market competition. Other cultures see the world differently and place greater value on cooperation.

Capra (1996) explains that the crucial differences centre on choices of objectives and responsibilities. The competitive paradigm aims at growth, counts the depletion of the Earth's resources as growth and sees all growth as desirable; the cooperative

Table 1.2: Improvements in performance from using more highly developed forms of partnering in the UK

	Construction	
	Cost	Time
Traditional approaches	100	100
First generation partnering	70	60
Second generation partnering	60	50
Third generation partnering	50	20

Source: Bennett and Jayes (1998).

paradigm takes sustainability as its over-riding objective. Sustainability means satisfying our needs without reducing the chances and choices available to our children and grandchildren. The competitive paradigm regards individuals as ultimately responsible here and now for themselves and their immediate families; the cooperative paradigm recognizes that we all have a shared responsibility for each other and for future generations.

The two views result in different theories about the structure of organizations. The competitive paradigm sees organizations as primarily hierarchical. It accepts that there are lateral linkages but sees them as working within a hierarchical framework. Managers are viewed as responsible for providing leadership, instilling values, deciding strategy and generally controlling their subordinates. Within this competitive paradigm, separate organizations compete with each other. Individuals prefer their organizations to succeed but believe never-the-less that they can be individually successful even if their organization fails. This gives pre-eminence to what Capra calls the self-assertive tendency in human nature.

The alternative view is based on what Capra calls the integrative tendency in humans. This paradigm sees organizations as richly interconnected networks working in cooperation with other organizations to form widespread networks. The internal structure of organizations is also made up of networks interacting with each other. All the networks interact in ways that are guided by feedback loops so that they form self-organizing systems. The values that guide decisions are grounded in sustainability. Within these complex systems, leadership changes depending on circumstances. Strategy emerges from interdependent decisions made throughout the organization's networks. There is wide recognition that an organization's success and the success of its individual members is interdependent. An important part of ensuring long-term success is exchanging information about performance with other similar organizations. This provides the basis for benchmarks that add a competitive spur to the work.

Neither the self-assertive nor the integrative tendency is intrinsically good or bad. What is good is a balance between the two. What is bad is to push either tendency to the exclusion of the other. There is a need to achieve a balance between female and male, south and north, east and west, art and science, command and communication, and cooperation and competition.

The right balance in any particular situation depends on the values of the people making the choices. Over the past 200 years too much emphasis has been placed on male values, the interests of the Northern hemisphere and, within that, to Western nations. Science has been given too much respect and the benefits of competition have been consistently over-stated. The results are deeply damaging to communities and the environment throughout most of our richly interconnected world.

The kind of construction that has resulted, and the methods used to produce it, suffer from the same fundamental weaknesses. All too often individuals and firms within the industry have taken a passive role in key decisions and allowed others to determine the form of its products and the nature of its processes. Inevitably, the resulting construction industries are weak.

No major construction industry has, as yet, managed to combine high levels of control and flexibility in a way that offers customers clear choices of value, performance, quality and service. Leading manufacturing industries have responded to Japanese methods by adopting lean production and partnering. The reasons are well described in Construction Task Force (1998; commonly known as the Egan Report) which uses experience from these other industries to set the targets shown in *Table 1.3* for the UK construction industry to improve its performance. The Egan Report goes further and uses experience from other industries that have adopted lean production and partnering, as well as research into leading edge practice in the UK construction industry, to recommend actions aimed at achieving the targets.

Table 1.3: Targets set for the UK construction industry in the Egan Report

	Improvement per year (%)
Capital cost reduction	10
Construction time reduction	10
Completion on time and budget	20
Defects on handover reduction	20
Reportable accidents reduction	20
Productivity increase	10
Turnover and profits increase	10

Source: Construction Task Force (1998).

One of the research reports drawn on by Egan is Bennett and Jayes (1998). This work describes the experience of leading firms using partnering in the UK construction industry and so provides some indication of the kind of industry that will result from Egan's recommendations. At their best, firms using partnering are managing to combine the efficiency of Japanese construction with the creativity of the best UK practice. Bennett and Jayes (1998) describe case studies of these important developments and draw carefully researched conclusions about the practical implications. The case studies include measurements of the improvements in costs and times being achieved. These show that as firms use more highly developed forms of partnering, which move closer to the methods implied by the new paradigm, their performance continues to improve. *Table 1.2* gives the most significant of the results which are based on careful measurements of the costs and times of comparable projects over what Bennett and Jayes call three generations of partnering.

Table 1.2 suggests that the construction industry can achieve the tough targets set by the Egan Report. To do so requires managers to see the construction industry, including its interactions with customers, in terms of networks in which the interactions behave broadly in the way illustrated in *Figure 1.8*. This will lead them to recognize that the industry's full potential depends on well motivated teams being empowered to cooperate within richly interconnected networks that, because they are guided by feedback loops, have the ability to self-organize.

Over and above that fundamental approach, managers need to guide teams within the overall pattern shown in *Figure 1.9*. This envisages the mainstream of the industry's work being sufficiently stable and predictable to justify concentrating on high levels of efficiency and quality. It requires managers to remain alert to surprises at the margins of mainstream work and to recognize that, on occasions, they will face unexpected and fundamental change. In readiness for these chaotic conditions, managers need to encourage new-stream work by undertaking research and development (R&D) projects or by responding to unusual customer demands. The new-stream projects will concentrate on creativity and innovation in developing a pool of new solutions that may provide the best response to some future change in markets or technologies. Indeed, managers responsible for construction should be involved in the forces shaping these major changes so that construction's interests and its skills and knowledge are taken into account.

Managers responsible for construction need to design organization structures and processes that fit these patterns of work. Only in this way will they provide controlled efficiency, be sufficiently flexible to deal with change, and ensure that the industry has a range of potential answers to problems as and when they arise. To achieve all this, managers need to adopt the new paradigm.

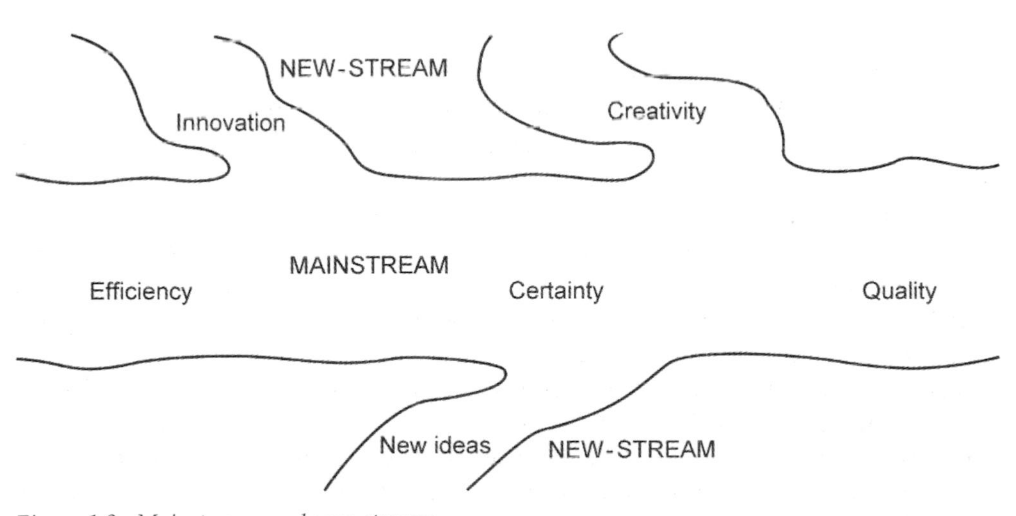

Figure 1.9: Mainstreams and new-streams

The next two chapters deal with the major forces of competition and cooperation. All too often these are seen as being in conflict. The great political divides are usually seen in terms of a right-wing belief in competition and a left-wing belief in cooperation. Construction needs a third way. The new paradigm balances competition and cooperation to provide a framework that helps managers to find long-term, sustainable answers to customers' demands in ways that deal with rapidly changing markets and technologies.

Competition

2

Market forces

Competition is widely seen as an effective spur to improve performance. This is clearly the case in sport where players try harder in competitive matches than in practice games. The sudden-death nature of knock-out competitions drives players to extraordinary efforts that are difficult to produce week after week in a league competition. International matches against countries regarded as close rivals or even enemies cause players to try harder than they do in routine matches for their club.

These same effects are seen in work situations where competition is widely used to motivate workers. Rewards are given to the team with the best quality record, or the best results in customer surveys, or the highest productivity. Firms compete by starting price wars with the aim of winning an extra share of the market for particular products or services. International free trade agreements aim at increasing competition, even though each country seeks to protect its own producers.

There is considerable agreement amongst economists that greed, the accumulation of personal wealth, is a basic human instinct. As is the instinct to trade. Together they make markets and competition inevitable features of human activity. The great strength of markets is that they provide information about the needs of communities, or at least the needs that people spend money to satisfy.

Economic theory begins with the concept of perfect markets, which means that resources are allocated so no one can become better off without someone else becoming worse off. In economic terms this is known as 'Pareto optimal' and in modern management jargon 'no win:win agreements are possible'. This remarkable result is achieved by what Adam Smith (1776) famously called the invisible hand of market forces. In a perfect market, the invisible hand gathers information about customers, suppliers and prices, and combines them in a Pareto optimal way. The condition necessary for a perfect market is that firms are in perfect competition. This means that all firms are too small to alter the market price by themselves, they have free entry and exit from the market and customers have perfect information about the goods and prices on offer.

These conditions are rarely met in practice, which is one reason why the costs and benefits of competition within free markets provide the subject for much political debate. On the one hand, liberals see government interference in markets as a threat to individual liberty or as restrictions on production. Social democrats, on the other hand, believe that governments should intervene in many areas in the interests of social justice and equality. Liberty and equality are desirable outcomes but, as the great rallying call of the French Revolution recognizes, achieving both requires a sense of community. Without a sense of community, liberty pushed too far creates huge inequalities and social divisions. Similarly, if equality is imposed, fundamental liberties are lost. There needs to be a sense of balance in all decisions about liberty and equality based on the interests of the community affected by the outcomes.

The collapse of the communist Eastern bloc headed by Soviet Russia at the end of the 1980s was widely heralded as proving the superiority of capitalism and competition in governing complex economies once and for all. Certainly the flow of economic migrants was overwhelmingly from the communist East to the capitalist West. The common view was that people had chosen free market capitalism and voted with their feet. They were attracted by the obviously greater wealth in the West and by the freedom that this gives most people. It was widely accepted that competition provided the most effective spur for greater production, and economic growth was seen as a realistic aim of government policy in almost every country.

Certainly these views are implicit in the established work methods of the construction industries in most countries. Firms are selected to undertake construction projects and the prices for their work are established by competition. Professional education and training very largely accept without question that the best way to select contractors is to invite competitive tenders from competent firms. There is no doubt that competition used in this way serves to drive down prices.

There are, however, growing concerns about the costs of relying on market forces. These concerns are evident at all levels including negotiations over international trade agreements, political debates within individual countries and amongst managers in many industries. International competition forces developing countries into debt and makes it more likely that their people will be subjected to starvation, poor health and worse. Competition within national economies rewards the rich more than the poor. Free markets tend to widen social divides which, in turn, brings many undesirable consequences including increased crime, family breakdowns and disaffected teenagers.

The costs in the construction industry may not be so dramatic but never-the-less they are significant; Latham (1994) describes some of them. They include poor quality work, late completions and an adversarial culture throughout the industry. Major causes of these problems arise from contractors' tactics as they struggle to earn a profit from prices established by competition. They dispute the adequacy of consultants' design information or the application of contract conditions. This leads to claims for additional money from the customer which can all too easily escalate into expensive litigation. Similarly, contractors use various tricks to force subcontractors into accepting unfair deals. Open competition and Dutch auctions are used to find subcontractors desperate for the work at almost any price. Then disputes are manufactured about whether the subcontractor's work is complete, to justify paying them late or not paying them at all. The ensuing negotiations often provide the basis for yet more claims against the customer. So, despite initially low contract prices established by competition, customers are often faced with extra costs during their projects even though they have not altered their requirements in any significant way. The important consequences are that many potential customers actively avoid new construction work because of the hassle it involves. Much-needed development is not undertaken and the UK has a weak construction industry that struggles to survive with low and variable demand, and very low profit margins.

Yet market-based competition provides undoubted benefits in motivating people to try harder. Lower prices attract customers and so provide a spur for producers to become more efficient. The central problem is that the benefits are not distributed equally, so there are winners and losers. Luttwak (1998) provides an interesting analysis of free market capitalism in the USA and concludes that it works because of a distinctive legal system and Calvinism. The legal system allows consumers, suppliers, lesser competitors and government agencies to use the law to contain the overpowering strength of big business. Calvinistic values persuade losers to accept sharp inequalities and winners to give much of their extreme wealth to charity. These restraints are necessary to enable free market capitalism to retain its credibility and Luttwak recognizes that they rarely exist outside the USA.

Moreover, capitalism works in the USA primarily because the USA is rich in natural resources. Justification for the US approach depends on the consumption of non renewable resources being counted as economic growth. This allows what is the most wasteful society on Earth to delude itself into believing that it is efficient. As Henderson (1993) shows, each unit of GDP in the USA consumes 2.5 times as much energy as the equivalent production in Europe or Japan. Yet the US approach is widely seen as a model for others to follow. Certainly it has strengths in encouraging enterprise but it also has serious flaws which mean that even this the richest country on Earth fails to provide a decent life for all its children. Market forces condemn 25 per cent of the USA's children to grow up in poverty in a society where the majority enjoy obscene levels of consumption.

This is the key point: market forces deliver a combination of benefits for some and serious social problems for others. The proportion of haves and have-nots varies enormously under different political regimes. In all cases the rich have to deal with the poor by accepting high taxes and generous welfare provisions; or by policing increasingly violent streets and turning their homes into fortresses. Wealth is redistributed by tax collectors and bureaucracies or by angry young men at the point of a knife or a gun. The costs involved in both approaches have steadily increased as the world has become more crowded. This has made it more urgent to find ways of providing the benefits of competition that support, rather than divide communities.

Beneficial competition

Beneficial competition unifies communities. It motivates people to beat the current norms, to improve on their own performance. It drives firms to invest in education, training, innovation and research. They use the information provided by competition to improve the service and value they provide for customers. They work at developing their workers' abilities and providing them with more effective support.

Competition provides information, the way we respond to it determines whether it is beneficial. For example, competition has allowed developed nations to distort the economies of poor countries by persuading them to grow cash crops instead of their own basic food. The resulting cash has all too often not found its way to those growing the food, and market forces eventually reduce those already poor to starvation. Competition should have been used in the long-term interests of poor countries by helping them to improve their ability to grow their own food and by providing education to develop new skills. Then gradually they can carve out secure markets for products in which they have some natural advantage. The same is true in construction. Weaker firms should be encouraged to concentrate on basic skills and then helped to improve these skills to high levels. Given this secure basis, multi-skilling can be added to provide a robust basis for the firms to develop their own futures.

It is our values and aims that determine whether competition is good or bad. Competition is bad when it drives firms into narrow cost cutting, down-sizing, making people redundant and subjecting them to unreasonable stress. Competition is good when it motivates people to improve on their own previous performance in ways that do not damage others.

Even when our motives are good, our actions may have some bad effects. So we need feedback to tell us when bad results begin to appear so we can change our actions to ones that we expect will achieve our good purposes without the bad effects. As long as competition is used for good purposes in the long-term interests of whole communities, and those making decisions use feedback to avoid bad effects, competition is beneficial.

Finding ways of using competition for beneficial purposes is crucial for the future of mankind. It is important that construction is involved in the search, not least because building sustainable communities will, in many parts of the world, make large demands on the industry.

Traditional competition

Traditional methods of managing construction projects use competition in a limited way that serves no one's interests very well. Customers are faced with decisions about procurement approaches which have little to do with the issues that determine value for them. A few of the competing approaches are illustrated in *Figure 2.1* but there are many others.

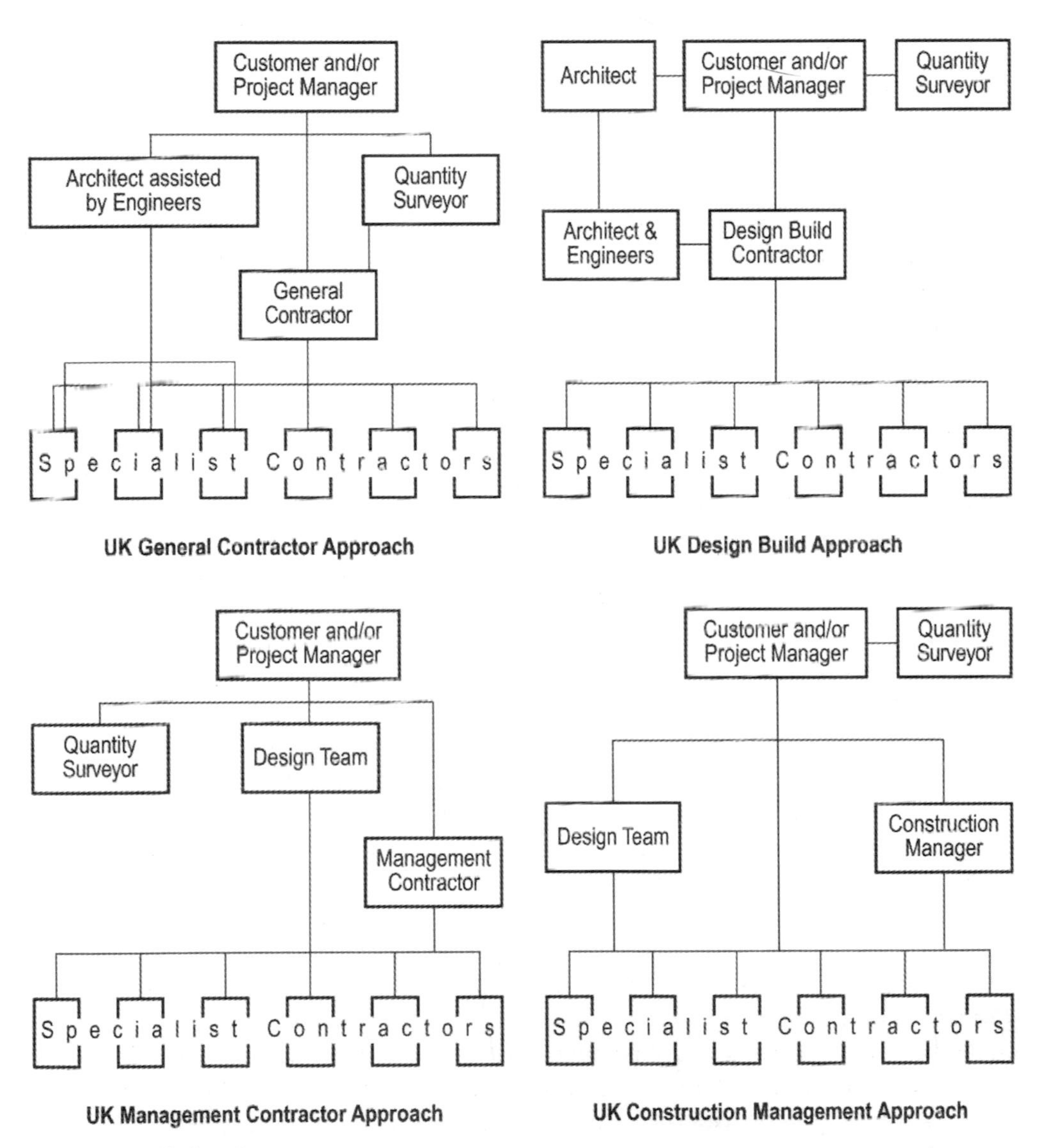

Figure 2.1: Choices for customers

Probably the least satisfactory but never-the-less common approach is for customers to appoint design consultants based on relatively casual recommendations. Fee arrangements are determined on a basis that may or may not involve an element of competitive bidding but which provides little if any incentive for the designers to search for the best value from the customer's point of view. Equally there is nothing to assure the customer that the designs produced by the consultants are even close to the best available. Then competitive tenders are used to establish the lowest price for the defined work and identify which amongst a group of main contractors, usually recommended by the design consultants, will undertake it at that price. The consultants keep the appointed contractor from contact with the customer and seek to blame the contractor for anything that goes wrong during the project. Contractors often win work by bidding a low price in the expectation that they will find sufficient mistakes in the design information provided by the consultants to justify claims for additional payments. Conflict is almost inevitable; all standard forms of contract in use in the industry include well developed procedures for dealing with disputes. Most of the direct construction work is subcontracted to specialist firms, often on the basis of dubious forms of competitive tendering which ensure that adversarial attitudes permeate every level of the industry. Customers eventually discover that they have only very weak assurances of getting a good product or being charged what they had assumed was the agreed price.

The adversarial attitudes and expectations of conflict mean that information is withheld in case it turns out to be useful to the firm that owns it in some future dispute. Misunderstandings are common and there is little or no sense of trust between firms. Contractor's market information is generally limited to identifying potential customers, discovering which have plans to invite competitive tenders and seeking to get onto the tender list. This approach gives customers no objective information to help them decide whether or not to invest in a new construction facility. Consultants work on the assumption that every project is unique and so see no possibility of providing customers with well researched, objective information about the choices available.

Alternative approaches deal with some of the worst aspects of these traditional methods. One approach is to add an additional consultant to manage the whole process on behalf of the customer. Project management and construction management are common examples of this alternative. Another common approach is to appoint one firm to take single point responsibility for the whole process as a design build contractor. Management contracting combines aspects of the management consultant and single point approaches. Many options have emerged within each of these main alternatives so, in total, there are a wide range of approaches. This tends to confuse rather than inform customers. Whilst the alternatives can provide greater certainty than traditional methods, they remain embedded in a competitive paradigm which does little to help customers make well informed choices about new construction before they are committed to significant expenditure.

There are parts of the construction industry that take a more proactive approach to fostering demand. In particular, development companies undertake extensive marketing, looking for opportunities where new facilities will be attractive to customers. Leading developers specialize in particular types of buildings and have well developed ranges of products and services aimed at categories of customers. Individual houses, apartments, small industrial buildings, shopping centres and offices are all provided in this way. At their best, developers provide customers with excellent information about the products and services they offer and the terms and conditions on which they are available. However, there are no widely understood standards for describing the various kinds of facilities and for measuring the developer's performance. So it remains for customers to discover what various competing developers can offer and then make their own comparisons of the value provided as best they can. They are considerably handicapped in so doing by advice from architects and engineers who believe that all projects should be seen as opportunities to explore new and creative solutions.

As far as the traditional construction industry is concerned, development companies are seen as customers and treated just like any other customer. The market information which guides the development companies' decisions has no influence on the approach used by construction firms. Construction firms compete for development work on the same basis as any new project. Having won a contract for development work, they organize their activities to maximize their profit, if necessary at the expense of the development company. There are some important exceptions where development companies have insisted on a different approach and have taken initiatives that help construction adopt a less competitive approach. These are however exceptions and, generally, traditional construction firms concentrate on organizing their work to achieve the biggest profit from a contract based on competitive tender.

So established traditional methods force firms in the industry to concentrate on internal issues and leave customers badly informed about the benefits that new construction can provide. Customers have little choice but to rely on price competition that offers no guarantees of good quality or value. This does not lead to beneficial competition. The key to using competition to benefit customers and the industry is to produce and publish good information that customers can use to demand more from the industry in ways that help and encourage efficient firms.

Customer information

The industry itself should take responsibility for ensuring that customers have all they need to make well informed decisions about new construction. This will benefit customers but it will benefit the industry itself even more because well informed customers make sure they work with firms that earn sufficient profits to invest in developing products and services that provide good value. Also, good customer information will encourage more potential customers to invest in new construction, so demand will be higher. Not only that, good customer information can be used to stabilize the level of demand and avoid the peaks and troughs that make management in the construction industry very difficult. The overall effects would be to make construction a more profitable and less risky business.

So there are powerful incentives to provide good customer information and this has never been easier. Modern information technology provides every encouragement to collect data, analyse it and communicate the resulting information to whoever needs it. The main barrier here is that effective use of these new technologies will require the industry to abandon old habits of secrecy and be open in exchanging information with competitors, customers and suppliers. Apart from the need to be open, the major practical issue for the construction industry is to define what information is needed by customers.

The industry and its customers face distinct situations which require different kinds of customer information. *Figure 2.2* shows the main stages of the decision tree facing customers and indicates the role that customer information plays in the key choices. The first decision point involves customers deciding, on the basis of information about the industry's mainstream products and services, whether a good solution to their requirements already exists. If it does, then the mainstream answer can be adopted. If there is no obviously acceptable mainstream solution, customers need to select a process that will help them to establish the criteria and value system that acceptable answers must meet. Once they have a better understanding of their needs, customers should look again at information about mainstream answers to see if any of them now provide a good answer. Failing that, customers need information about new-stream processes that describe how they will approach the task of solving the problem or exploiting the opportunity and delivering the solution for a price lower than the value to the customer.

Customers also need crystal-clear statements of the terms and conditions on which the industry is prepared to do business. These should tell customers the costs and risks involved in buying a new building or other facility and the time necessary for design and construction processes before their new facility will be available for use. Customers also need to know how much of their own time and resources will be taken up in making decisions throughout the whole process.

This information is most effective when it is produced by organizations that are independent of any industry firms. In the UK, the Consumers' Association produces a magazine called *Which?* that reports the results of very careful tests of consumer products and services. It provides information about the features of competing products, describes their performance in use, evaluates the initial and running costs and arrives at firm conclusions about the best buys. The reports are independent of the producers of the products and, because of this, are widely regarded as reliable guides.

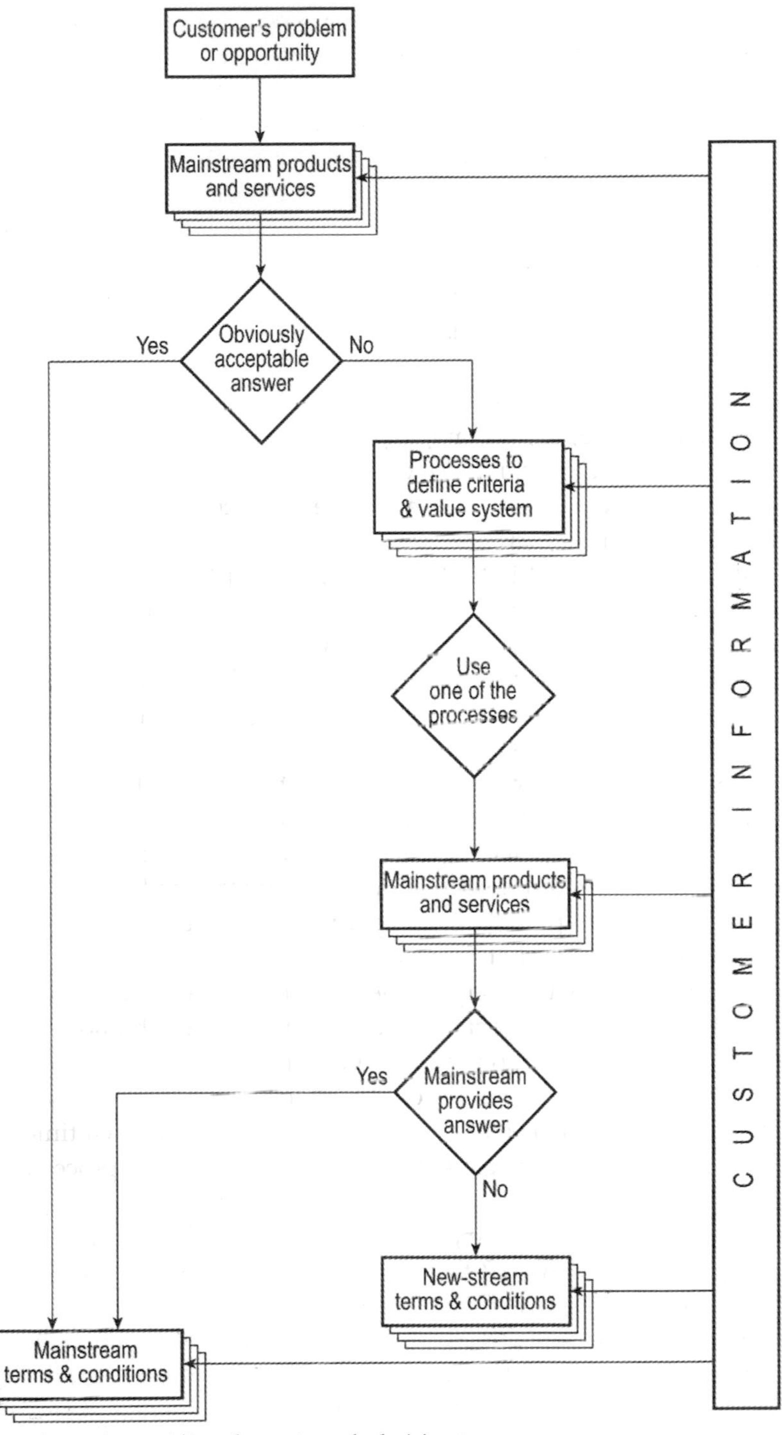

Figure 2.2: Customer information guiding the customer's decision tree

Construction has largely failed to ensure that its customers have similarly thorough and independent information. Yet, much relevant information exists. For example, leading house builders in the UK have made some progress by working with the Building Research Establishment to devise objective measures of the performance of the houses they produce. Similarly studies in the UK, USA and Japan under the name ORBIT, described in Duffy and Henney (1989), have established good measures of the functions and performances of office buildings. Also there are well established design guides to many common building types that provide a mass of detailed information about their functions and performance requirements.

However, generally the industry has taken the view that its products are unique and so it is not possible to devise performance measures that would help customers. This perception comes from concentrating on differences and ignoring the many features of buildings and other constructed products that are very similar as far as customers are concerned. In no small part the cause of the industry's narrow view of its work comes from seeing it as very complicated.

It is of course true that modern buildings and many other constructed facilities are very complicated products but they are made up of many common features. Taking buildings as an example, many of the spaces they provide are common. The most obvious evidence for this is that they have well known names. Thus terms such as kitchen, toilet, dining room, bedroom, office, entrance hall, etc., conjure up clear images of what should be provided. Customers and construction professionals know the essential features of these spaces.

In addition to these common spaces, most buildings include some which have more specialized requirements. Information about these requirements is less widely distributed but firms who specialize in producing these spaces, and their regular customers, have well developed measures of the performance criteria necessary for the spaces to fulfil their intended functions. There are very few spaces that are truly unique and so the vast majority of building projects could and should be guided by good customer information.

So the physical complexity of the industry's products can be dealt with and good customer information provided. However, the primary cause of the widespread perception that the industry's work is complicated arises from the great variety of interests that have to be dealt with during construction projects. Chapter 6 describes the basic processes that the industry works through and identifies the many different interests involved in construction. Even in the earliest stages of considering a new construction project these are likely to include, in addition to the customer, financial institutions, local and regional authorities, public utilities, neighbours, special interest groups and the users of the new facility. The users may well include several distinct groups, all with a real interest in any new development. These typically include people who will work in the new facility, their senior managers, the customer's customers who will visit the new facility, suppliers who need to make deliveries, and the facilities managers who will run the new facility once it is complete.

So there are many interests to be considered from the earliest stage of a project. It is of course a complex business to deal with all these different interests. When this organizational complexity is overlain with competitive attitudes, it dominates thinking about construction. The complexity is real and Chapter 6 describes the processes needed to manage it. However, the complexity of interests faced by construction teams should be an internal matter for the industry, and should remain an internal matter because there is no long-term benefit in forcing customers to think about construction as a complicated business. This is exactly the situation facing the car industry. Modern cars are wonderfully sophisticated and complex products but customers expect clear information about the issues involved in owning, driving, running and maintaining them. Construction customers should be offered similarly clear, simple options and be given help in choosing amongst them. They should not be forced to worry about all the separate interests involved in designing and carrying out the work.

Market research

The way to provide clear, simple options and so produce good customer information is to concentrate on the factors that determine the decision whether or not to go ahead with a new construction project. This is what market research does; it identifies the features of products and services that are important to this key buying decision. *Figure 2.3* illustrates the main steps in using market research to provide good customer information. These steps should form an ongoing process, guided by feedback from customers using the information, to make decisions about new construction.

The first step is to ask properly selected samples of customers about their experience with particular types of facilities and the important differences they experienced. This provides a basis for identifying the key features and the criteria customers use in valuing them. When this research is undertaken thoroughly, it identifies and takes account of all the interests that influence customers' key buying decisions.

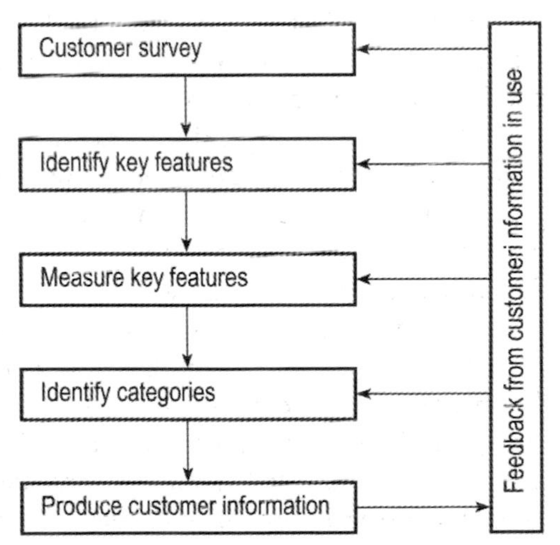

Figure 2.3: Using market research to produce customer information

Having identified the key features and criteria, the next step is to devise ways of measuring the features based on the customers' criteria using methods that are as objective as possible. This stage will draw on a wide range of skills and knowledge. Many of the features will be measured using precise, scientific methods. These may involve laboratories, purpose designed test facilities and measurements of facilities in use. Not all features can be measured objectively, for example aesthetic issues have a subjective content. This does not mean that customers cannot be given good information about them, but it means they need to be measured using subjective techniques and customers need to be guided in using the results.

Having decided how the key features should be measured, the next step is to arrange for an independent body to measure the products and services produced by competing construction organizations. In other fields the results often identify several distinct performance and quality categories which are usually given distinctive names such as luxury, executive, sports, family and economy. To some extent this kind of classification exists in parts of the construction industry. Houses and offices are described in terms that indicate the quality standards that customers can expect. Duffy and Henney (1989) propose a set of descriptive names for office buildings that include: big city high spec., city fringe medium spec., highly serviced towers, smaller city high identity, smaller traditional spec., high tech. and utility.

However they are described, the categories of performance and quality need to be related to differences in price so customers can identify the best value from their own point of view. Given good information, customers are well able to do this. For example, Green (1994) describes how value management uses workshops at which the key members of a construction project team help the customer identify his or her own value tree and relate costs to each of its branches. In this case, the process requires a workshop to bring all the main interests together so the customer can base his decisions on good information. The point is that, given good information, customers can decide what represents value.

Workshops are currently necessary to bring together specialists capable of dealing with all aspects of complicated construction products. However, workshops are expensive and the powerful tools provided by modern information technology offer cheaper alternatives. Given that the basic information describing value and costs exists, current data manipulation systems are well able to turn the specific needs of individual customers into statements of value and cost. Flanagan *et al.* (1998) describe systems that do this and indeed evaluate customers' needs in terms of time, quality, and many other factors relevant to a serious comparison of alternative products and services. These sophisticated systems have the potential to help customers explore many options before the decision is taken to go ahead with a new project. *The Economist* (1999) describes how the Internet is already being used to exploit this potential in other industries.

As described earlier in this chapter, the customer information needed to drive evaluation systems is most effective when it is produced by organizations seen as independent and therefore trusted. The Consumers' Association was quoted as a good example of a body undertaking this role for many consumer products and services in the UK. Another good example is the specialized magazines that publish independent reviews of cars. These provide a mass of measured information about costs and performance, detailed test reports by experienced drivers and surveys of customer satisfaction. It would be foolish to buy a car without using the information these magazines provide. Yet the traditional construction industry leaves its customers with no choice but to act in just such a foolish manner. The absence of good customer information forces them to consider construction projects with little understanding of what it is reasonable to demand of the industry or what they should expect if they decide to go ahead with a new project. The absence of well informed customers inevitably results in a weak, poorly performing construction industry.

The link between well informed customers able to set targets that demand ever higher levels of performance and value, and the emergence of world-class firms is evident in many other industries. In the absence of such demanding customers, construction is able to continue working to its inefficient and unsatisfactory norms. Market research provides the tools to identify the information that customers need in order to make good decisions about investing in new construction. It is in the industry's own best interests to make sure this information is produced and published.

The key role of the resulting customer information is to establish targets that will motivate construction firms to improve their performance. Bennett and Jayes (1998) describe several cases where customers with programmes of regular, similar construction projects set targets based on carefully analysed information about their own projects. The examples include the Esso Oil Company establishing targets for their service stations across Europe, Sainsbury's establishing targets for its superstores, the British Airports Authority setting targets for several kinds of facilities including runways, warehouses and office buildings, and the Rover Group setting targets for new buildings associated with the design and testing of cars. *Table 1.3* indicates the scale of improvements possible.

The improvements result from more than just setting targets but in all these cases the use of challenging targets that are accepted as being achievable is an important element in the industry's success. Certainly these cases provide strong support for the idea that targets set by well informed customers provide a competitive spur that is more effective than making decisions on the basis of the lowest price submitted in competitive tenders.

The point is that good customer information provides for real competition between construction organizations, based on the issues that are important to customers. It allows customers to make a broad evaluation of the options in terms of quality and value. When price is crucial then good customer information helps customers invite competitive tenders and evaluate them on the basis of their own quality and value criteria. The existence of broad categories of products and well established measures of their key features provide the basis for beneficial competition. Whether competitive tenders are used or not, the important outcome from providing customers with good information is clear targets for construction organizations to deliver the products and services customers want.

Construction's use of targets

Having provided customers with information that enables them to be confident in setting targets, the challenge for the construction industry is to use the targets to improve its performance. This requires the paradigm shift described in Chapter 1. The best way to understand why this is so is to review current approaches to using targets in control systems and identify why a new approach is needed.

The use of targets is a recurrent topic in management literature. Galbraith (1973) provides an influential description of the role of targets in managing organizations. It was written within the old management paradigm and sees targets as an effective means of coordinating the work of independent teams. This is based on the widely held management view that independent work is efficient and Galbraith's insight was that it can be coordinated by individual teams working to carefully set targets.

Bennett (1991) describes how Galbraith's approach applies to construction projects just as much as to other management situations, but argues that it is difficult to set targets that coordinate independent work without some degree of waste. The waste arises because, if tough targets are set which demand consistently high levels of efficiency, some teams will fail to meet their targets. This is likely to have a disruptive effect on the work of other teams and require additional management effort to re-establish coordinated work. Bennett concludes that for targets to coordinate independent work, some degree of slack or waste must be allowed so that all teams can achieve their targets. Slack targets for cost may be justified if, for example, speed is important. In such cases, targets can be set to enable several teams to work simultaneously, safe in the knowledge that their work will fit together provided all the targets are achieved. Slack targets may also be justified if the cost of the waste likely to result is less than the costs of the additional management required to deal with failures to meet tight targets. So, in practice, slack targets are widely used because they provide coordination without making unreasonably large demands on management. As a further benefit they allow individual teams to enjoy the feeling of success that comes from meeting targets.

Bennett (1991) regards this approach as a good first step and suggests that once managers have achieved coordinated work and well motivated teams, they should progressively set tougher targets which require more efficient performance. He suggests that this is how efficient organizations are developed. By seeking better performance, year after year, organizations gradually become world-class. This plausible argument ignores the fact that the competitive paradigm uses targets to keep teams separate and independent of each other.

The fundamental problem with this is that targets have to be comprehensive if they are to allow completely independent work. As Peters and Waterman (1982) famously state 'more than two objectives is no objectives', by which they mean that if, for example, a team has targets for costs, time, quality, safety and customer satisfaction, situations will arise where choices and trade-offs between them need to be made. Gradually every target is brought into question and changed to accommodate new circumstances so that eventually none of the original targets survive. Peters and Waterman correctly conclude that targets are effective only if they deal with just one or two key objectives. This means that targets cannot be both effective and provide enough information to define the work of independent teams sufficiently closely to coordinate it.

Bennett (1991) attempts to deal with this dichotomy by distinguishing between two kinds of target. First, are constraints which must be met in order for the end product to function satisfactorily and for it to be produced at costs and times that meet the customer's requirements. Second, are those targets which are to be the focus of an organization's efforts to improve its own performance. The second category is what Peters and Waterman see as crucial in motivating teams to search for ways of improving performance.

Bennett illustrates the distinction between constraints and key objectives by using the example of an engineer designing a structural steel frame. Typically structural engineers have a set of targets which include the nature of the loads transmitted onto the frame by other elements, the loads transmitted by the frame onto the foundations, a maximum size for columns and beams, a minimum column spacing, the maximum deflection in beams, a maximum cost, the timing of design, manufacture and construction processes, and the planned life of the frame. In addition, it may be decided on a particular project to concentrate on reducing the total weight of the structural frame by 10 per cent of current norms and halving the normal deflection in beams. These targets for improvement are examples of Peters and Waterman's key objectives. The rest are constraints which must be met by the design. As long as the constraints are based on the engineer's own performance norms, he can concentrate most of his efforts on the two key objectives.

So independent teams can have key objectives to provide a focus for improved performance provided they work within constraints that are sufficiently comprehensive to coordinate their independent work. As well as the performance criteria for individual teams' work, constraints should deal with all the technological and organizational interactions between teams. For all practical purposes this is impossible for a new design. It would be necessary virtually to design, manufacture and construct the complete product in order to establish totally comprehensive and realistic constraints. Consequently, the use of constraints as a means of coordination is feasible only in respect of standardized projects. In other situations an evolutionary, step-by-step approach to identifying constraints must be adopted.

Control systems

A step-by-step approach is implicit in the design control systems used by leading consultants. Good examples are described in Gray *et al.* (1994). However, a flexible, step-by-step approach is more explicitly evident in cost and time control systems.

Well developed cost control systems typically assume an agreed budget which is subdivided into elements to enable design and management proposals to be reviewed one at a time, as they emerge. Normal practice is to include a separately identified contingency element which provides a degree of slack in the overall target. The allocation of the contingency element in response to cost problems is managed with the aim of encouraging designers and managers to search for better answers. In practice it is mainly used to provide a safety net to cope with difficult problems. When main elements are subdivided to provide more detailed cost targets, good practice uses the same approach of setting tight targets for sub-elements plus a contingency to provide a degree of slack. The approach continues throughout the cost control process as finer and finer cost targets are set to whatever level of detail is deemed sensible for a particular project. By this means design and management decisions can be reviewed one at a time, so that separate professionals can work independently yet the overall cost consequences of their decisions are coordinated in accordance with the budget agreed with the customer.

Similarly, well developed methods exist to coordinate the planning of design, manufacture and construction programmes. They, too, provide a structure of an overall completion date, medium-term targets and short-term targets. Bennett (1991) suggests that the industry would do well to fit projects into one year and within that into months, weeks and days. He argues that any construction work that requires more than one year should be divided into smaller projects which take twelve months. Be that as it may, within each of the separate time elements, experienced managers employ a combination of tight individual targets and a degree of slack to provide a safety net. This improves their chances of achieving the overall agreed programme without having to deal with major crises.

Feedback

The final point to deal with in this review of construction's use of targets is the important role of feedback. All controlled systems, including construction projects, depend on feedback. As *Figure 2.4* shows, feedback enables teams to exercise control on the basis of their actual outputs rather than on their planned performance.

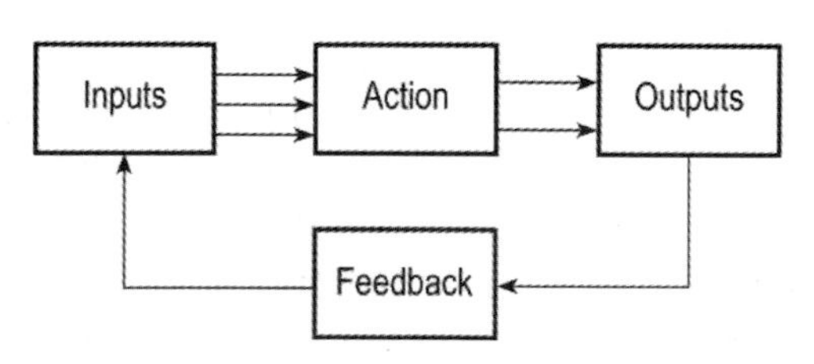

Figure 2.4: Basic control system

Feedback consists of knowledge about the effect of actions on a system's environment. Control comes from the system using its feedback to guide future actions towards agreed targets. This is achieved by comparing measurements of its performance with the targets and, depending on the outcome of that comparison, deciding to continue with the same actions because they have produced an acceptable result or, when the feedback shows that the results deviate significantly from the targets, changing the actions to bring the feedback into the range of acceptable answers.

A problem which construction project managers need to take into account is that there are considerable differences in the time-scale of construction feedback. There is a great deal of immediate feedback. The craftsman working with a material obtains second-by-second feedback on the effect of his actions. He can see the effect of his actions on the material and he can feel the effect through his hands as he works. However, feedback on the performance in use of the part he is making will necessarily take time to become available. Similarly, an architect sketching the first designs for a building may have to wait years before his actions are joined in a feedback loop by knowledge about the appearance of the actual building.

Construction teams are generally concerned with short-term feedback; that is, feedback which arrives in time for it to influence their behaviour while they are still working on the project. Longer-term feedback is valuable in that it provides knowledge which helps future projects; however, traditionally, it is of less interest than short-term feedback in managing an individual project. The fact that some important parts of construction project feedback are long term has traditionally justified managers in making decisions without all the relevant information. This is one important reason why many decisions are very poor.

The fact that feedback on any one project is less than complete is an important reason for construction to use well established answers wherever possible. Standards and procedures built up on the basis of feedback over many projects provide a robust basis for new projects. Indeed, society at large recognizes this and imposes regulations of various kinds, including planning and building regulations to control what may and may not be built. These provide important constraints which teams need to take into account.

However, where they depart from well established answers, teams need to give particular attention to ensuring that they have reliable feedback to provide early warnings of problems. This is crucial in helping the teams to know where they should concentrate their efforts. Yet all too often the task of finding new answers is so demanding that the need for feedback is ignored.

Target-driven competitivity

The overall effects are that, at present, construction's use of targets in control systems is a mixture of procedures based on the old competitive paradigm and developments that anticipate what is needed for the future. These need to be taken further and the first step in describing what is required is to identify the principles that should guide the industry's procedures to deliver competitivity. In other words, we need to identify what is necessary for the targets set by customers to be used as effective drivers for improved performance. The way that cost targets should be used illustrates the principles. These are summarized in *Table 2.1*.

Table 2.1: Principles of effective procedures

1	Overall targets reflect value to the customer
2	Construction organizations get fair returns
3	Team targets fit within overall targets
4	Team targets are agreed and achievable
5	Teams are jointly responsible for meeting all targets
6	Targets aim at high performance
7	Teams have everything needed to meet their targets
8	Teams have feedback about their own performance
9	Problems are tackled quickly and decisively
10	Successes are celebrated and rewarded
11	Failures are opportunities to find better answers

It was established earlier in this chapter that giving customers objective information about the options open to them empowers them to set challenging but achievable targets. These should guide the work of the construction organizations involved in producing the required products and services so that the overall target cost reflects the value the new facility will provide for the customer. So the first principle is that the value of construction to the customer provides the overall budget.

The second principle is that the procedures provide a fair return for the construction organizations involved. Again, cost targets provide a good illustration. The budget is allocated by first agreeing a fair profit for each construction organization involved. It is sensible at the same time to calculate an allocation to cover a fair contribution to fixed overheads for each of the organizations.

The third principle is that the overall target is an absolute limit in calculating targets for construction. In terms of costs this is achieved by deducting the allocations for fair profits and fixed overheads from the overall budget and what remains becomes the cost target for construction. This is divided into targets for the direct costs of undertaking the work for each separate construction organization.

The fourth principle is that these individual targets should be achievable. A good way of ensuring this is for teams to establish their own targets. There is a great deal of evidence from practice that targets set by the team who will carry out the work are far more likely to be achieved than those which are imposed. This is true even when the targets are set at the same level. The sense of commitment engendered when a team feels that it has set its own targets motivates them to try very hard to achieve them. The benefits go beyond this because teams tend to set tough targets for themselves when they are given measurements of their own previous performance and believable evidence that direct competitors are achieving better performance.

The fifth principle is that everyone involved has a responsibility for ensuring that all targets are met. A commonly advocated approach is for each team to be given information about how their particular targets fit into the overall objectives. Then all the teams should be asked to look for and discuss ideas that will help them meet the targets. Each good idea should be discussed until there is agreement on a set of targets, consistent with the organization's overall targets, which all the teams are committed to achieving.

The sixth principle guides these discussions, it is that targets should aim high by assuming that everything will go well. So, in cost terms, no contingency allowances are included because they allow teams to settle for weak answers. In the same spirit no overall contingency is allowed. In other words the sum of the tough individual targets, the agreed profits and fixed overhead allocations equals the customer's budget.

The seventh principle is that teams are provided with everything they need to meet their targets. This means that only very occasionally will a team fail and so meeting targets becomes an essential part of their integrity and honour. Given this, once a team has accepted a target and has the resources needed to meet it, they concentrate on doing their best work rather than looking for excuses for failure.

Then, as work proceeds, the eighth principle is that teams should have feedback about their own performance related to their individual targets.

The ninth principle is that any problems that threaten the achievement of a target should be tackled quickly and decisively. Teams should act to solve problems as soon as they become aware of them. When a key target is in danger of being missed, this must be treated as a crisis and clear, effective action taken quickly by all the teams involved to get the work back on target.

The tenth principle is that success should always be celebrated. When targets are met, teams should get immediate, tangible rewards so they feel their efforts are valued.

The final principle is that any failure should be taken very seriously. Not in any sense to allocate blame but to take the opportunity to find a better, more robust answer. Some organizations celebrate failures to emphasize that they provide important learning opportunities. As long they are used to make real improvements, this seems a good idea.

Benchmarking

Benchmarking is consistent with these principles which are summarized in *Table 2.1*. Benchmarking is a tried and tested process that uses carefully devised targets to continuously improve business performance. It is based on researching the extent to which others carry out the same or similar processes more efficiently, identifying how they do this and selecting features of their methods that can be used to improve performance. The improvements are then put into practice and the results measured. Benchmarking is an on-going process and so the measured results provide a starting point for the next cycle of improvement.

Pickrell *et al.* (1997) provide good evidence that benchmarking can help construction teams undertake objective analyses of their work. It makes them aware of best practice and helps them understand how they can apply its key features to their own work. In this way benchmarking engages people in a systematic search for ways to improve their own performance.

Benchmarking consists of a series of practical yet scientifically based techniques developed in the leading manufacturing industries of the USA and Japan. The benchmarking model described in Pickrell *et al.* (1997) provides a carefully researched and tested approach tailored to the needs and circumstances of the UK construction industry. As *Figure 2.5* shows, it consists of a series of distinct steps which begin when an organization recognizes the need to improve performance. This may be triggered by seeing benchmarks describing the performance achieved by others. It may arise from internal concerns over profit levels, the share price falling or threats from new forms of competition. The first step should be an internal workshop to review whatever it is that has caused the organization to recognize the need for change. The output from the workshop should be an agreed statement of the need for change.

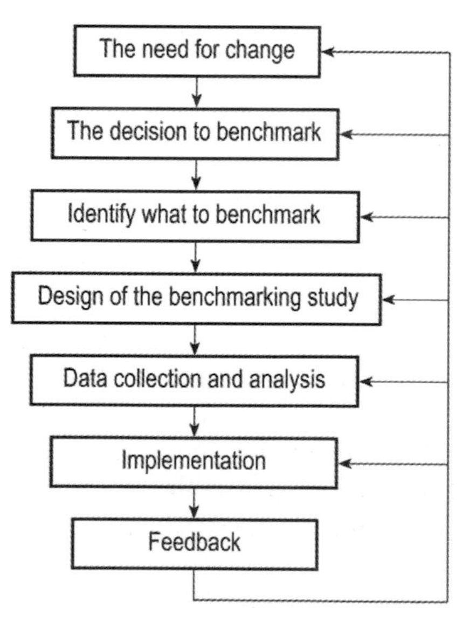

Figure 2.5: Benchmarking process

The second step is the decision to use benchmarking. Inevitably there are costs in using benchmarking. These include training people in the process of benchmarking and the costs of workshops and facilitators. Pickrell *et al.* (1997) report that there are many examples in construction where the benefits far outweigh these costs, never-the-less the costs need to be considered and balanced against the benefits in making the decision to benchmark.

The third step is identifying what to benchmark. Discussions resulting in the need for change statement will help in identifying which process should be benchmarked. In general the aim should be to start with the process where there appears to be the greatest need for improvement and the biggest benefits if a better answer is found. The ease with which change can be introduced, especially the likely attitudes of the people involved, should also be considered in deciding which process to benchmark.

The next step is to assemble existing descriptions of the structures and methods used in the selected process, including the measures of performance currently used by management. These descriptions and measures provide a starting point for a process map which is best prepared at a workshop by the staff responsible for the process. The aim is to identify the core areas of the process that add value for customers. For each of these core areas, the factors that are critical to its success should be identified. Then these success factors should be prioritized on the basis of their contribution to the organization's overall success.

The workshop should agree how the top priority success factors can best be measured to reflect the organization's performance. These critical success factors and strategic metrics, together with the process maps of the targeted areas, form the preliminary framework for a benchmarking study. The next main step is to design the study. Benchmarking essentially compares strategic metrics from different organizations to identify differences in performance and so helps organizations to decide where to look for improvements. The benchmarking partners can be other divisions of the organization, competitors or non-competitor organizations undertaking the same function or generically the same activity. It is sensible to have five or six partners to provide a range of approaches. To ensure their full commitment, the potential partners have to be convinced that there is a real probability of mutual benefits in undertaking a benchmarking study.

The partners' level of commitment becomes apparent as the details of the study are agreed. A critical stage is often when the partners attempt to agree the detailed measurements to be used in comparing their performance. These must be defined and agreed by all parties, as must the precise data needed, the collection methods, the role and responsibilities of each partner and the degree of confidentiality that applies. Putting these agreements into practice and collecting the data needed to compare performance makes significant work for all parties. Agreeing the time and resources required to carry out this stage of the benchmarking study is often the real test of partners' commitment.

Having agreed these key details and maintained the commitment of all the parties, the staff likely to be affected by changes aimed at improved performance should be told what is being done. They should have an opportunity to comment on the arrangements and to be involved in the study.

The next main step is the actual data collection and analysis. The rules of measurement for the tactical metrics should be tested in a pilot study to ensure that they are practical and provide robust measures of performance that are genuinely comparable.

Having completed a pilot study, the metrics and data collection methods to be used should be reconfirmed or altered to take account of the results. Then, as the performance of all the partners is measured, regular checks should be made to ensure consistent methods are being used and the measurements are truly comparable. Once all the data is assembled and checked carefully for ambiguities and gaps, it should be described in a formal written report that is available to all the parties involved in the study.

The next step is crucial in identifying scope for improvements. It consists of analysing the data to identify differences in performance between the partners. The most significant differences should be investigated to ensure that the data is accurate. Then the reasons for these differences must be identified by comparing the organization's own process analyses with similar descriptions of the best performing organizations.

Each significant performance gap should be reviewed to decide if it is critical to the organization's success. If it is, then a target and timetable for closing the gap should be set. Targets that will give quick and positive results help establish the benefits of benchmarking. The actions needed to achieve the target should be planned using the process analyses produced during the study.

The detailed targets and plans should be made available to all the parties involved in a formal report. It is particularly important that the report is communicated to those whose support will be needed and to those who will be affected by changes.

The planned actions are then put into effect. The action plans should provide for feedback by arranging for the resulting changes in performance to be measured and compared with the targets. It is also usual to include regular reviews of progress at which the parties discuss their successes and failures. Any shortfalls should be investigated and every possible action taken to achieve the targeted improvements.

Pickrell *et al.* (1997) finally recommend that the improved performance level achieved as a result of the benchmarking study should be embedded in a new target or benchmark which provides a starting point for further studies.

The approach recommended by Pickrell *et al.* was tested in benchmarking studies within the UK construction industry. The tests showed that amongst a group of five or six construction firms, all were best at some aspects of the subjects of the study and all could find improvements in their own performance.

These results encouraged contractors working for the same one major customer to undertake benchmarking studies. By working together, the contractors found they can all improve the service they provide for their common customer. Perhaps more importantly the measured benchmarks provide objective information for the customer about the contractors' performance. The point the contractors have recognized is that they are assured of the customer's work provided they continue to deliver demonstrably good value. So they are in competition, not with each other, but with all the other firms who would like to replace them as the customer's preferred contractors.

The recognition that they have a common interest in cooperating to become better at doing the customer's work than anyone else and their use of benchmarking has produced a stream of measured improvements in the performance of all the contractors. Bennett and Jayes (1998) report cases where it would now be very difficult for other firms to match the resulting performance levels. This is an excellent example of beneficial competition and how it should be balanced with cooperation in the interests of customers and construction organizations.

Cooperation 3

Effective behaviour

Cooperation is the most effective behaviour to adopt in all human relationships. It is only if others do not cooperate that competitive behaviour is sensible. This important idea has been a feature of human life for at least as long as records have survived. It is, however, only over the past two decades that the benefits of cooperation in business relationships have become widely accepted in the literature.

Cooperation means people working together on the basis of common interests. This means, for example, they take care to act in ways that do not damage each others' interests. When it is not possible to meet everyone's interests, they agree joint plans of action that aim at sharing the pain on a basis that all regard as fair.

The recent acceptance of the benefits of cooperation was triggered by its success in major Japanese manufacturing industries, notably the car industry. This has been the industry most characteristic of the twentieth century and is now dominated by management methods that depend on cooperative relationships. Womack *et al.* (1990) report a major research study of the world's car industries which shows Japan to have by far the most efficient. They found that Japanese car firms typically produce between 30 per cent and 60 per cent more cars per worker than their US or European competitors. Womack *et al.* explain how Japan's efficiency depends on long-term cooperative relationships between manufacturers, their suppliers and their customers. Bennett *et al.* (1987) describe similar levels of comparative efficiency based on using long-term cooperative relationships in Japan's construction industry. The most careful evaluation of the relative efficiency of various national construction industries is reported in Atkins *et al.* (1994). At purchasing power parity, as *Table 1.1* shows, Japan has the most efficient construction industry in the world, with comparative costs some 25 per cent to 30 per cent lower than those in the UK or USA.

Similar practical evidence of the benefits of cooperation has emerged in many other industries and has led to the publication of a great deal of advice on how cooperative relationships should be managed.

The book by Axelrod (1984) is absolutely key in the development of ideas about cooperation. It describes many natural and human situations in which cooperation has emerged and identifies the principles which explain why this happens. The essential and important message of Axelrod's work is that many situations can be modelled as the 'Prisoner's Dilemma'. This provides the basis for a number of effective management games which have two parties making independent decisions about whether to cooperate or compete. There are four outcomes, for which *Figure 3.1* shows a typical set of payoffs.

		Decision by A	
		Cooperate	Compete
Decision by B	Cooperate	Both get 3	A gets 5 B gets 0
	Compete	A gets 0 B gets 5	Both get 1

Figure 3.1: Payoffs in the 'Prisoner's Dilemma'

These payoffs produce outcomes similar to many real life situations: the players in a game interact an indefinite number of times so they do not know when the last interaction will take place; they remember how they both behaved in previous interactions; there is no way for the players to make enforceable threats or commitments; there is no way to be sure what the other party will do on the next move; there is no way to eliminate the other party or avoid the interaction; and finally, the players cannot change the payoffs.

The dilemma illustrated by games based on the payoffs shown in *Figure 3.1* is that, whilst a given community is better off if everyone cooperates, individuals can be better off by competing. However, faced with a party that competes, there is no benefit in continuing to cooperate. So competitive behaviour corrupts efforts at cooperating and causes everyone to compete. When this happens the whole community is worse off than if everyone cooperated.

The results from a game between two parties of ten rounds of decisions based on the payoffs in *Figure 3.1* serve to illustrate the point. If both parties cooperate on all ten rounds, they both score 30, making a total for the pair of 60. If one party competes and one cooperates on all ten rounds, the first party scores 50 and the second 0. The total for the community formed by the two of them is 50 which is less than when they both cooperated. However, it is extremely unlikely that the second party would continue to cooperate and score no points faced with a relentlessly competitive approach from the first party. Assuming the second party decides to stop cooperating after three rounds and compete for the remaining seven, it scores 7 and its opponent scores 22, making a total of 29. This is now significantly less than if they had both cooperated. The first party may feel good because it has won but it

		Decision by A	
		Cooperate	Compete
Decision by B	Cooperate	60	50
	Compete	50	20

Figure 3.2: Total points from ten rounds of 'Prisoner's Dilemma' using the payoffs from Figure 3.1

has less then it would have had from cooperating and the community is significantly worse off. *Figure 3.2* shows the total payoffs produced by ten rounds of the two parties making consistent decisions. The best outcome results when both cooperate. The strength of Axelrod's research is in showing that many natural and human situations fit this pattern of payoffs.

This provides Axelrod with a robust basis for the view that human communities are most successful when people expect each other to act cooperatively. This is not a soft, idealistic approach because successful communities punish competitive behaviour. However, they also forgive people when they return to cooperative behaviour. Axelrod calls this 'Tit For Tat'. Within games based on the Prisoner's Dilemma, 'Tit For Tat' means that you begin by cooperating on the first move and thereafter copy the other party's previous move, whatever it is. An effective variant is 'Tit For Two Tats' which means cooperating on the first two moves and continuing unless the other party acts competitively on two successive moves. This reflects a good principle in human relationships of giving people a second chance but not a third or a fourth or more.

The success of 'Tit For Tat' over more competitive strategies suggests that in human interactions the most effective strategy is to be nice, to retaliate against competitive behaviour but to forgive transgressors when they return to cooperative behaviour. It is also important that this strategy is absolutely clear so that people understand each other's intentions. This is because people generally give up on those who act randomly or create the impression that their cooperation cannot be relied upon.

For cooperation to evolve in real life situations the main requirement necessary is that individuals should have a sufficiently large chance of meeting again that they have a stake in future interactions. The likelihood of future interaction is crucial because if individuals come to doubt that they will meet again, they tend to seek their own individual interests and act competitively. This implies an understanding that, in situations that fit a one-move game based on the Prisoner's Dilemma, it pays to compete. Only in situations equivalent to an iterated Prisoner's Dilemma does it pay to cooperate.

The problem faced by all living things, modelled by the Prisoner's Dilemma, is that while the community as a whole benefits from cooperating, each individual can do better by exploiting the cooperative effort of others as long as they can avoid retaliation. This means that where interactions are random or are not repeated, behaviour will be competitive. The payoffs are less than in a cooperative community but entirely competitive communities survive because everyone gets some payoff from all interactions.

However, when the same individuals expect to meet more than once and they can remember previous interactions and at least some of the outcomes, the Prisoner's Dilemma applies and cooperation is the best strategy. This is important because Axelrod shows that the pattern of many biological systems fits the Prisoner's Dilemma and cooperation explains their behaviour. In other words the benefits sought by living things are more readily available to cooperating communities.

Once established, cooperative strategies resist being taken over by parties that compete so long as everyone retaliates quickly and forgives once the deviants return to cooperation. Even in a world of unconditional competition, the likelihood of future interaction can give rise to small clusters of individuals who cooperate based on reciprocity. In a world where many strategies are being tried, cooperation based on reciprocity can thrive. Cooperation, once established on the basis of reciprocity, wins over less cooperative strategies and therefore continues. In other words cooperation provides a robust and stable basis for a community, competition does not.

Cooperation is used by many higher-level animals. They have the skills needed to tailor their behaviour so they can reward cooperative behaviour and punish competitive behaviour. They are good at distinguishing between individuals, can remember the outcomes of previous interactions, estimate the probability of future interactions and determine the likely outcome of specific interactions. Given these skills, and humans have them in abundance, cooperation can emerge, even in competitive environments. At a fundamental level, pairs of related individuals, classically parent and child, instinctively calculate payoffs in terms of their joint effect and cooperate. Thus choice of behaviour depends on how others behave. Once cooperation exists, the key determinant of it continuing is reciprocity of cooperation; Axelrod describes many human situations where groups of individuals expect to meet again and cooperation spreads to the mutual benefit of them all.

Axelrod regards communities in which everyone competes as being in a primitive state which is evolutionarily stable. But, as the members evolve more advanced abilities, cooperation emerges by one of two mechanisms. First, as described above, kinship gives individuals a stake in each other's success. Second, clusters of cooperating mutants emerge in which individuals interact frequently and flourish more successfully than their competitive neighbours. Whichever of these mechanisms enable cooperation to emerge, 'Tit For Tat' provides the community as a whole with the largest benefits.

Cooperation strengthens individuals, relationships and organizations at all levels. Tough-minded cooperation based on careful discussions of everyone's interests is empowering. It allows people to behave reasonably because they know they will be treated reasonably. It follows that an absolutely key part of tough-minded cooperation is identifying and responding to those who behave selfishly. Anyone acting in ways that damage a community of which they are a part should be stopped.

An important issue is the extent of the communities to which individuals should be expected to give this kind of loyalty. Many different views are held but the more mature tend to take an increasingly wide view of the communities for which they should take some responsibility. In an important analysis of what is happening on a world scale Cooper (1996) describes three kinds of countries. Those characterized by internal chaos are pre-modern. Those that have established effective internal law and order and believe strongly in state sovereignty and non-interference by one country in another's internal affairs are modern. They behave competitively in international relationships and tend to be dangerous to their neighbours. Third, are post-modern countries, most of which are in Europe, who have outgrown their hang-ups over sovereignty, encourage mutual interference in each other's domestic affairs and invite constraints and surveillance in military and other matters that threaten neighbours.

Post-modern countries accept that communities extend well beyond national boundaries, as evidenced by such organizations as EU, NATO, IMF and the World Bank. Business increasingly reflects this wider view in taking account of how its activities impact the environment and the communities in which it operates. The emerging definition of communities is that they include all those affected by actions. This view leads of course to the idea of subsidiarity, that decisions should be taken democratically at the level of community that includes all those affected.

These developments at least implicitly accept that in the long-run all levels of community are better off cooperating. However, many individuals in these communities still need to be persuaded of this. In the context of construction this requires investments in training and robust audits. It may also mean deciding not to work with some individuals or firms because they are determined to act competitively in pursuit of their own narrow interests. However, as Cooper's post-modern countries and the vast majority of leading manufacturing firms demonstrate, the general trend is towards the mature state of interdependence.

Mature adults

The practical actions that put cooperation into effect in business-to-business relationships are widely called partnering. As Bennett and Jayes (1998) describe, partnering is being applied at the leading edge of the UK construction industry, mainly in response to demands from major customers for greater efficiency, faster completions and better quality. The approach they found being adopted in practice is consistent with basic principles of effective behaviour, the most fundamental of which is that the greatest success in human activities results from mature adults building long-term, cooperative relationships. There is wide agreement that people must become independent adults before they can successfully form interdependent relationships.

Covey (1989) describes the habits adopted by successful people. One of the bases of this influential book is a review of the last 200 years of literature about how to be successful. Covey discovered that during the first 150 years of this period, the character ethic was seen as the foundation of success. This taught that there are basic principles of effective living and people can only be successful if they integrate these principles into the way they behave. The principles include such things as integrity, humility, fidelity, temperance, courage, justice, patience, industry, simplicity and modesty. It focuses on the importance of people taking responsibility for themselves and accepting that success depends on what happens inside ourselves.

Over the last fifty years the character ethic has disappeared from the literature and has been replaced by descriptions of a personality ethic. This sees success as resulting from public image, attitudes and behaviours. It concentrates on skills and techniques that win friends and influence people. It causes people to see problems as resulting from factors outside of themselves. It causes them to seek to blame others rather than looking within themselves for the solutions to their problems.

Covey takes the view that this has been a mistake and, in so far as the literature reflects the actions of human communities, is the cause of much misery. It leaves people with empty lives and makes it difficult for them to form effective relationships. So Covey (1989) describes actions that help people develop their character and be more truly successful. Covey's set of actions, which he calls seven habits, provides the basis for a growing maturity as people move from the childhood state of dependence, first to independence and then to interdependence. The first three actions, which help move from dependence to independence, are: be proactive, begin with the end in mind, and put first things first.

Be proactive

The first action is based on the view that success depends on what happens inside ourselves. Everyone has the freedom to choose how they react to events. Mature adults do not blame other people, circumstances, conditions or other external factors for their behaviour. They recognize that their success results from their own conscious choices. It is not what happens to us but how we respond to it that determines how successful we are. The same point is made by Axelrod (1984) in describing the strategies that lead to success. He advises people to concentrate on their own success and not to be envious. In particular people should consider whether they are doing at least as well as any other person would in their situation, not how well they are doing compared with other people with whom they interact.

Proactive people use phrases such as: I can choose, I prefer, I will act, I control my own feelings, and similar indications of a positive attitude. Reactive people blame other people, circumstances or the stars for their situation.

Covey provides a useful classification of problems into those involving our own behaviour over which we have direct control; those involving other people's behaviour over which we have indirect control; and those which we can do nothing about, such as past mistakes or situational realities.

Having classified their problems in this way, effective people concentrate on the problems they can deal with by being proactive. If they set a goal, they work towards it. If they make a promise they keep it. They do not criticize others but try to make their actions a model of correct behaviour. In this way their behaviour becomes part of the solution, not part of the problem. Finally, for the problems over which they have no control, they smile and gracefully accept the situation.

So the first step towards becoming a mature adult is to accept that everyone is responsible for their own success. It means accepting responsibility for our own situation and not blaming others for problems. This is encapsulated in the Japanese saying: 'whoever first sees a problem, owns it'. This means that when anyone sees a problem they take responsibility for making sure that it is solved. If they are competent to resolve the matter, they do so. If not, they find a person who is competent to deal with it and ensure that it is tackled. In either case they do not ignore problems or dump them onto someone else and walk away.

Begin with the end in mind

Covey's second action is developing a clear view of what is really important. This should come from thinking carefully about the purpose of your life. Covey suggests forming an idea of how you want you life to be viewed when it is ended. What would you want your family, friends and colleagues to say about you when you are dead? Having created a mental picture of how you want your life to be seen, it should guide your actions. At the start of each day ask yourself how your actions will contribute towards creating the kind of life you want your family, friends and colleagues to remember.

Creating the vision of your life or the life of an organization for which you are responsible is an act of leadership. It means deciding on the broad direction, deciding what business to be in, deciding what outcomes are desirable, deciding on the purpose of our lives.

Covey suggests that people base their lives on various centres, including their spouse, family, money, work, possessions, pleasures, friends, enemies, a church or themselves. These centres are ultimately weak and result in a limited, narrow life in which people swing from one centre to another. As a result they have no consistent sense of direction, no persistent wisdom, no reliable power and little sense of personal worth.

The proper centre for human lives is principles. The correct principles for any individual come from within themselves on the basis of their own conscience. However, Kidder (1999) describes interesting research suggesting that there is a widely accepted ethic based on eight common values. These are accepted by people in many different cultures and so provide, at the very least, a good starting point.

The eight values identified by Kidder's research are love, which people saw as encompassing compassion, caring and empathy. Next is truth, including honesty and integrity. The next two are freedom and a sense of unity or solidarity. The fifth common value is tolerance, so that majorities do not stamp out minorities and diversity can flourish. Sixth is fairness, which includes equity and justice. Seventh is responsibility in the sense of being accountable for ourselves. The final value is respect for others, the environment and the diversity of species.

Love, truth, freedom, integrity, tolerance, fairness, responsibility and respect provide a powerful checklist for people as they work out their own principles, ethics or values. However they are established, Covey suggests that individuals should express these deep beliefs in a personal mission statement based on thinking carefully about what is really important to them. Personal mission statements should be developed, reviewed and improved at regular intervals. They give structure, commitment, exhilaration and freedom to a person's life.

Personal mission statements should deal with each of the main roles considered important by the individual. These often include spouse, boss, parent, colleague, friend, charity worker, political activist and more. The statement should identify the goals the person wants to achieve in each of these areas of their life.

Having written a mission statement, you should live it. One effective help in doing this is to spend time visualizing yourself behaving in accordance with your mission statement in difficult situations. This mental rehearsal has been found to be effective in many situations. All world-class athletes routinely use visualization to prepare for difficult moments in competition. Effective managers do the same, they mentally rehearse difficult meetings, the stages of a negotiation, the way they will deal with a difficult subordinate, and so on.

The importance of developing a clear view about what is really important is echoed by Fisher and Ury (1981) who found that successful negotiators spend time deciding what is really important to them so that at the beginning of a negotiation they can describe their own needs, wishes, concerns and fears.

Put first things first

Having decided what is important, effective people concentrate on the activities of greatest worth to them. Covey (1989) found that effective people put this third action into practice by organizing each day around priorities concerned with achieving the results and goals of their personal mission statement.

Day-to-day activities can be classified on the basis of their importance and urgency. Effective people do not spend time on unimportant things which use up their time but do not help achieve their personal mission. They deal with important, urgent things quickly to make space for the important, non-urgent things that contribute to their long-term success. Covey gives as examples of important, long-term actions: building personal capabilities, building relationships, recognizing new opportunities, planning long-term actions and recreation. *Figure 3.3* summarizes this approach to making better use of our time.

	Not Urgent	**Urgent**
Unimportant	Do not spend time on	
Important	Set time aside to deal with	Deal with now

Figure 3.3: Time management matrix

It is difficult to stick to a policy of putting first things first and many people at home, work and play allow themselves to get bogged down by trivial things. Many find comfort and safety in doing routine, unimportant things which waste their lives. Mature adults know what they want to achieve in all aspects of their lives and concentrate their efforts on actions directed towards these aspirations.

Covey recommends planning our use of time a week ahead so as to concentrate on important, non-urgent things. He suggests the plan should consider all the important roles played by the person producing it. The first step is to identify two or three goals for each important role. Then time is scheduled, allowing for daily adapting to deal with unexpected events but all within the framework of the week's goals. Covey's wide experience of helping people use this approach shows that the more closely the week's timetable is tied to a wider framework of correct principles embodied in a personal mission statement, the greater the increase in effectiveness it provides.

Putting the week's timetable into effect is difficult because urgent problems and tempting opportunities arise unexpectedly. Never-the-less any decision to allow the timetable to be altered should be guided by the person's own principles and mission statement. It is much easier to act on this advice if people learn to say no gracefully when problems intrude or opportunities provide temptations that would cut into what is truly important from the individual's own point of view. This is not selfish because the plan allows time for all the important relationships. The problem is that it is often easier to say yes.

One fundamental way of gaining time is to delegate. Effective delegation means you have less to do but you need to invest time at the outset to ensure that the other person understands exactly what is needed. This requires that you clearly define the desired results by helping the other person visualize them. You should tell them of any established guidelines, which should be few in number but which provide warnings of common failures and traps. You should identify the resources the person can draw on to do the work. You should clearly define the standard by which the results of their work will be evaluated and describe the consequences that will flow from success and failure. Then you must do what you promised, spend more time on important, non-urgent things and leave the other person to learn from taking responsibility.

Cooperative interdependence

The greatest success in human affairs is achieved by moving beyond the independence that comes from being a mature adult to the higher state of cooperative interdependence. Carlisle and Parker (1989) use the important concepts of ultimate customers and supply chains to explain why cooperation provides the most successful approach in business relationships. Their essential idea is that cooperation between contracting parties in any one supply chain is a far more powerful strategy for making both parties more profitable than any adversarial approach yet devised.

The approach begins by recognizing that when ultimate customers buy specific products, they are deciding which supply chain to support. The ultimate customer either uses all the parties in a given supply chain or none of them. Therefore, all buyer:seller relationships have a mutual interest in the ultimate customer's business which nourishes them both. They also have a mutual interest in making their supply chain more effective than those of competitors. It follows that in commercial negotiations within any one supply chain, the other party is not the adversary. The real adversary for both parties is competition from other supply chains.

Carlisle and Parker show that the largest untapped opportunities for competitive advantage lie at the interfaces between firms in any supply chain. It is therefore good business sense for both parties to concentrate on enlarging the benefits of their interdependency, not on increasing their share of the benefits. This is commonly expressed as making the pie bigger, not fighting over how big a slice each gets. However, this focus must recognize that there has to be a sense of fairness about the division of the pie.

An important part of making a bigger pie is that all parties to any relationship are prepared to sacrifice their own short-term interests to foster their joint long-term success. Therefore all commercial deals should be seen as the potential start of a long-term relationship in which the parties want to cooperate in finding better ways of working together over time.

Carlisle and Parker describe a mass of evidence which shows that the most successful supply chains are based on long-term relationships which grow out of relationships where, at every significant interface between the parties, their needs are met in an ongoing way. What matters is that *needs* are met, not wants or wishes. So, over time, a successful supply chain meets more and more of the needs of the firms that make up the chain.

The needs that have to be met are whatever the people involved perceive as important. People's perceived needs change over time, not least because they exist in a hierarchy that reflects a growing maturity. Carlisle and Parker describe a simplified hierarchy of needs in the following terms:

1 Existence needs: all forms of psychological and natural desires, including money.
2 Social needs: relationships with other people, including the need for individual worth to be recognized sufficiently to build a sense of self-esteem.
3 Growth needs: include the need for individuals to be creative and achieve the full potential of their talents.

The long-term success of any interaction depends on ensuring that the people involved satisfy their lower-level needs and can develop to satisfy higher-level needs. In this way people involved in successful relationships are rewarded with steadily growing personal maturity.

So there is considerable agreement that cooperation is the best strategy for many commercial relationships. There is further agreement on the principles which should guide the resulting interdependencies. Although they are described in various terms, there are essentially three fundamental principles which Bennett and Jayes (1998) see as the essential basis of partnering in the construction industry. They call the three principles mutual objectives, decision making and continuous improvement, and this terminology is used here.

Mutual objectives

The first essential basis for cooperative interdependence is to agree mutual objectives. Covey (1989) expresses this in terms of the fact that successful people enter into only win:win agreements. This recognizes that win:win is the only viable long-term strategy. All other options result in adversarial relationships in which ultimately everyone loses. Covey defines win:win as a frame of mind that seeks mutual benefit in human interactions. It is based on the idea that there is a better and higher way than my way or your way if only we take the trouble to look for it. The better way is our way in which we both win by working towards agreed mutual objectives.

In searching for win:win situations, an abundance mentality is extremely helpful. This means accepting that there is sufficient for everyone to have all they need if only they work together to create it. Therefore people should keep talking until a win:win solution is found. If this is impossible, the only viable option is to accept that no worthwhile agreement can be found. This is why it is important to have thought out an alternative course of action. The freedom of being able to accept the possibility of not doing a deal is incredibly energizing and encourages people to be really creative in looking for win:win solutions.

Table 3.1 summarizes the steps suggested by Covey (1989) in looking for win:win agreements. These steps closely reflect the approach proposed by Fisher and Ury (1981) based on their finding that successful negotiators begin by focusing on interests rather than on fixed, predetermined answers.

Table 3.1: Steps in finding win:win agreements

1 See the problem from the other's point of view so that each party can express the other person's needs and concerns as well as they do themselves
2 Identify the key needs and concerns of both parties
3 Determine what results would constitute an acceptable solution
4 Identify options to achieve those results
5 Evaluate the most promising options
6 Select the best option and try it out
7 Evaluate the results against the parties' key needs and concerns
8 Repeat any or all of the previous steps to provide even bigger wins

Source: Covey (1989).

Focusing on interests requires negotiators to talk about their own interests in specific, practical terms. Good negotiators talk in detail about what they want for the future. In this way both parties can look for solutions that satisfy the other party's interests. So, an essential basis for good agreements is that both parties are clear and open about their own interests.

The next step is to work with the other party in generating a range of possible solutions. Good negotiators use creative techniques such as brainstorming, looking at problems from the point of view of different professions, or thinking of newspaper headlines that might be used to describe the outcome. Working together creatively can produce a relaxed frame of mind that opens negotiations to a search for answers which provide mutual benefits. In contrast, unsuccessful negotiators quickly focus on specific solutions. The problem with this is that if a fixed position is genuinely unacceptable, it merely creates head-on confrontation. Focusing on interests provides the best chance of finding mutually acceptable answers.

The resulting win:win agreements should be written down and signed by all the parties involved. They should at least state the desired results in terms of what is to be done and by when; the constraints within which the results are to be achieved; the resources available; the system of accountability, in terms of the standards and timing of evaluations; and the actions that will flow from evaluations.

Decision making

The second principle of cooperative interdependence is to agree how decisions are to be made. The agreement should provide an organizational framework for making decisions and guide the behaviour of the people involved. The broad aims should be that few problems arise and those that do are dealt with in ways that do not threaten the relationship.

Organizational framework

The first step in establishing a decision making framework that forms part of a cooperative interaction is to ensure that senior management really does want to work long-term with the other party. Then, within all the organizations involved, there needs to be a consistent policy of ensuring in all interactions that the other party's interests are taken into account. A good way of doing this is to appoint named individuals from each firm to deal with specific issues. This brings groups together frequently which, because they expect to meet again in future interactions, gives them a reason to cooperate.

Carlisle and Parker (1989) put these ideas into a formal organizational framework that identifies all the interfaces between the parties. Carlisle and Parker call these the touch points and argue that the perceptions and relationships at each one need to be consciously managed. As *Figure 3.4* shows, both firms should identify the manager responsible for each touch point and so create pairs of managers responsible for sustaining one specific aspect of the relationship. It is important to ensure that the touch point managers have the skills to determine and deal with the underlying needs of the other party.

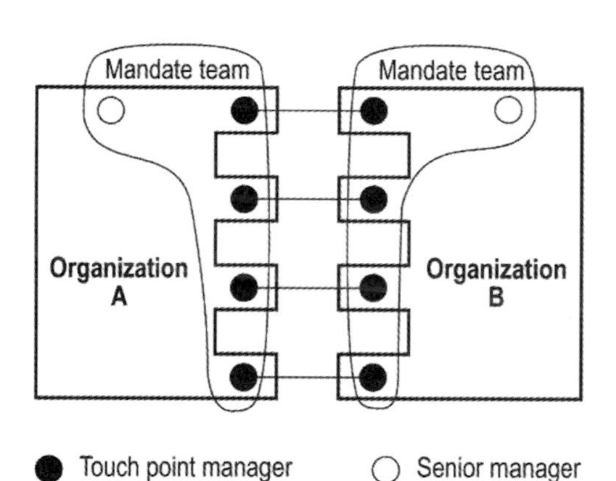

Figure 3.4: Organizational framework for cooperative interaction

At each touch point there should be an audit process which measures success or failure in terms of the parties' perceptions of the extent to which their needs are being met. The key criterion should be each party's perception of the other manager's willingness to help meet their needs. Questions should be asked regularly of the pairs of managers to monitor the state of each touch point. The results should be used to find ways of improving the relationship.

In addition, each firm should set up a mandate team which has the authority to take key decisions about the relationship. The team should include all the touch point managers who, between them, should ensure that all key interests are represented. The mandate team's role is to make key decisions about objectives and methods and make sure there is consistent effort towards sustaining the relationship. Carlisle and Parker also suggest that each party should appoint a senior person to ensure that their firm keeps all its promises. Without this, promises may get forgotten, trust cannot develop and relationships breakdown.

Organizational processes

The way in which major problems should be tackled within these organizational frameworks is described by Heirs and Pehrson (1982) in terms of the four-stage model shown in *Figure 3.5*.

The first necessary stage is to define the question to be answered. This should be stated as clearly as possible, including identifying the criteria that satisfactory answers need to satisfy. These should take account of the overall mutual objectives of the parties.

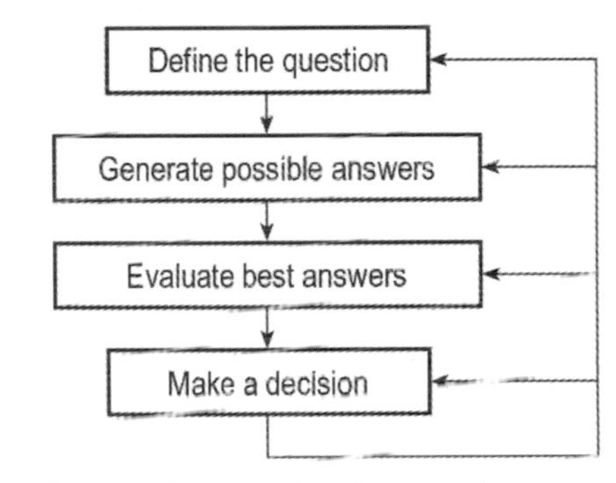

Figure 3.5: Decision making framework

Second, a number of possible answers should be proposed. There are many creative techniques that can be used at this stage. Designers are likely to use highly creative design methods. Construction managers may well use brainstorming and similar techniques designed to free groups from thinking of existing answers and help them find new ideas. The aim of the second stage of problem solving is to identify a number of possible solutions.

Third, the two or three most promising answers are evaluated. They should each be judged against the criteria identified during the first stage. The strengths and weaknesses of each should be listed, particularly taking account of the consequences on other elements of the work. It is important to be seen to be fair at this stage. So, for example, if a variable has to be valued, this is best done by reference to market value, reference to a previous similar case, judgement by an independent professional, normal practice, or similar objective criteria to provide a fair standard or a fair procedure.

Finally, a decision should be made. This may be to select one of the answers reviewed at stage three. It may be to adopt a combination of elements drawn from several of the potential answers. It may be a decision to evaluate more of the answers identified at stage two or to search for more potential answers. When none of these options appear attractive, the original question may be reviewed. Alternatively, a task force may be set up to search for an answer. Indeed, in major construction projects it is common in the early stages to set the same problem for two or three task forces to work on simultaneously. Answers are then presented by each task force in front of the others so that subsequent discussion can provide a more rounded, deeper understanding of the likely source of a good answer.

It is wise to adopt the four-stage problem-solving approach in formal procedures. Heirs and Pehrson suggest it is universally effective, being based on the way the human brain itself searches for answers. Thus, the four-stage approach applies also to each of the stages. For example, in defining the question at stage one, it is sensible to propose a number of possible formulations of the problem to be solved. The best should then be evaluated in terms of their clarity, the extent to which they encourage creativity and similar criteria. Only then should the question to be tackled be selected. Several ways of generating options should then be considered before deciding exactly how to tackle the next stage, and so on.

Individual behaviour

Having set up organizational frameworks to support effective decision making, attention should be given to ensuring that people behave cooperatively. Axelrod (1984) found that effective people are never the first to compete because they know that the consequences are likely to be to their disadvantage. They reciprocate both cooperation and competitive behaviour, otherwise people are tempted to try and take advantage of them. Also they do not try to be too clever. It is important that intentions are clear and are reflected in people's actions. Being too clever or using complicated reasoning does not pay, it is most likely to be seen as devious or random.

Carlisle and Parker (1989) suggest that successful decision making requires people to understand the other parties involved so that they can anticipate problems and work out solutions. They explain that solutions to problems must be grounded in whichever level of need people perceive as most pressing at any given time. In dealing with other people it is essential to take account of their perception of current needs, whether these are basic existence needs, social or growth needs. It is not sensible to make judgements about other people's perceptions of their own needs. Whichever need they perceive as most pressing is what will motivate them to accept solutions to problems.

Fisher and Ury (1981) also describe research which shows that good negotiators tackle people issues head-on. Good negotiators describe their own feelings and impressions without allocating blame. They encourage others to describe how they feel and why they view the problem that they are negotiating over in the way they do. They create opportunities for others to describe everything that is bothering them, so there are no remaining, hidden problems or concerns. They do not talk about the other parties' attitudes or motives. They do not fall into the trap of attributing their own motives and concerns to others because this nearly always leads to misunderstandings. They listen more than they talk. They ask questions to check their understanding of what the other party has said. They summarize and restate the other party's case in a positive light in order to build agreement. They check that the other party has understood what they themselves have said. Finally, they make sure that their own interests are clearly stated and understood.

All this practical advice is consistent with the judgement arrived at by Covey (1989) that communication is the most important skill in developing cooperative interdependent relationships. He concludes that the most neglected part of communication is understanding the other person. We tend to assume that other people experience the world in the same way that we do and assume that we understand them because we have experienced what they are going through. This assumption is nearly always wrong and leads to many misunderstandings. Because we try to understand other people based on our own motives and behaviour, we evaluate, probe, advise and interpret from our own frame of reference and experience. All of which builds a sense of resentment and alienation. Instead people should give time and attention to understanding the other person. This includes understanding what is really important to them and then making it important to you. It includes attending to the little kindnesses and courtesies that show you really understand their point of view.

It is also important to clarify all your expectations and bring them out into the open, especially in any new situation. Covey recognizes that this takes courage and time but it prevents misunderstanding and disappointment later. Building confidence in your own integrity is crucial to any effective relationship. It is important to remember that the way you talk about people who are not present tells the other parties how you will talk about them when they are not present. If you tell them secrets told to you by others, they will suspect that you will break their confidences, so they will not be open; this is why integrity is crucial to successful relationships. It means being totally honest in everything you do. It means keeping commitments and not making promises that you do not keep. An important part of this is to apologize sincerely when you behave badly towards another person.

Good decision making requires empathic listening. This means listening until we understand so well that we can rephrase what the other person has said in a way that reflects their feelings. This takes time but provides the only secure basis for effective decision making. You should continue listening until the other person is confident that you understand them so they can take the risk of trusting you. Once a person is understood, affirmed, validated and really appreciated, they will be willing to work on decisions in a wholehearted way. By taking the risk of really listening and trying to understand, you empower the other person and provide a basis for you to help them meet their needs.

Developing deep understanding may need practice in empathic listening. This can include role-playing the approach you should adopt in imagined difficult situations so as to train yourself to listen. Replaying situations in which you behaved badly can also help develop the habits of cooperative behaviour.

Knowing how to be understood is as important as understanding. This means taking account of the other person's concerns and describing your own needs and wants in their terms. Deeply understanding other people in this way opens the door to creative solutions. Different points of view cease to be stumbling blocks, instead they help people find better answers.

These cooperative behaviours are most effective when they operate within an overall decision making framework in which people first define the problem to be tackled and then work together to find an answer that takes account of all their interests. When a solution is found, effective people frequently allow the other party to take the credit for having suggested it. This helps them feel they own it and so they are more committed to making it work.

Continuous improvement

The third principle of interdependence is that people in successful relationships invest in the continuous improvement of their joint physical, spiritual, mental and social well-being. Covey (1989) sees this as dependent on teamwork which provides the synergy to make the whole greater than the sum of the parts. Synergy requires that you have the courage to describe your real needs and feelings, to understand the other person's real needs and feelings and then together search for an answer that gives you both more of what you need.

The essence of teamwork is to respect differences, to value them, to build on strengths and to compensate for weaknesses. Effective teamwork requires people to open their hearts and minds to new possibilities, new alternatives and new options. Many people are frightened by these highly creative activities, they want structure, certainty and predictability.

Teamwork is the development of creativity with other human beings. The teamwork principle recognizes that by valuing different points of view, we increase our knowledge of the world. By looking for the good in others, you enlarge yourself. By having the courage to be open, to express your ideas, your feelings and your experiences, you encourage others to be open also. In any discussion, if two people have the same opinion, one of them is redundant. We need to value different perceptions and try to understand them. Everyone should communicate with people who see things differently and value the differences. We can learn from people who see the world differently if we seek to understand and affirm their point of view. In doing this we need to remember that people do not see the world as it is but as they are. Two people can disagree yet both be right. Respecting differences makes it more likely that people will find a third alternative that provides a basis for continuous improvement.

Carlisle and Parker (1989) argue that teams need three key freedoms if they are to have the confidence to cooperate in looking for better ways of working:

1 *Freedom of thought*. This requires that teams have all the relevant information so that they can build up an accurate picture of the situation. Then all ideas can be considered.
2 *Freedom of feeling*. Everyone has the right to express their feelings about issues so that balanced decisions based on 'head and heart' can be made. Such decisions should carry authority throughout the organization.
3 *Freedom of action*. Once an action is agreed the team responsible for it should have full authority to carry it out.

Kaizen

Continuous incremental improvement, year after year, is one important basis of Japanese industrial success. Imai (1986) explains this is called 'Kaizen', which is achieved through procedures enabling Japanese firms to draw on the good ideas of teams throughout their whole workforce. Kaizen is a long-term process that encourages managers and workers to develop initiatives where the benefits will take time to emerge. Kaizen has been introduced into Western industries through training in process analysis, work measurement and basic statistical techniques. Kaizen is applied by teams in workshops that begin with the team's manager describing the organization's vision of improved products and better ways of working. This leads into discussions in which the team considers what they can improve, what level of improvement they can aim for, what stands in the way of improvement, what are the first steps they must take to achieve the target level of improvement, what help they need from others in order to make the improvement, and how the improvement will be measured.

Having understood the basic ideas of Kaizen, the starting point for applying them is to identify a problem to be tackled. The main criteria for selecting problems should be their impact on customers: those that inconvenience customers should be tackled first before dealing with internal issues.

Once a problem is identified, the process which gives rise to the problem is analysed in terms of the causes and effects that determine performance. This analysis identifies all the inputs and outputs as well as the stages in the process. *Figure 3.6* shows a typical pattern for the resulting process diagrams.

The next step is to measure the key features of the processes surrounding the problem. The measurements are used to help establish the extent and location of the problem. Then a measurable target is set for improving the existing situation. This should be an ambitious target set by the team responsible for the process in agreement with their manager.

Specific actions designed to achieve the planned improvement are then agreed, usually on the basis of intense discussions within the team. This stage may draw in experts or researchers from inside or outside the firms involved in the Kaizen study.

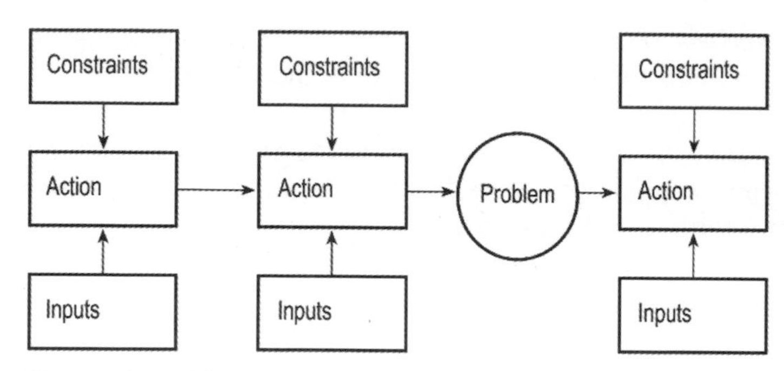

Figure 3.6: Problem analysis diagram

The aim is to find a robust improvement that the team feels confident about. Then the actions are put into practice.

The effects of the change are monitored by further measurements of the performance of the process. Hopefully the actions will produce their intended effects, in which case the team should look for another problem to solve. However, if the actions do not solve the original problem, the earlier steps should be reviewed to look for more effective actions. In either case the search for the next improvement, either in the same process or another one, continues.

It is normal for teams to measure their own improvements in performance and report their achievements twice a year to their managers so that everyone knows whether they are doing better this year than last and by how much. These reports form a key part of staff evaluations and decisions about rewards and promotions.

Japanese construction companies have annual Kaizen competitions, in which basic work teams describe problems that they have tackled and the improvements that they achieved. These competitions generate hundreds of ideas for improving working methods and are an interesting symbol of the contribution which Kaizen makes to the industry's efficiency.

There does, however, appear to be an unrealistic assumption in Japan's use of Kaizen that incremental improvement can continue for ever. Japan's financial problems in the late 1990s may well be caused in no small part by this faith in never-ending improvement. As Handy (1994) describes, all technologies have a life cycle which takes the general form of a sigmoid curve. *Figure 3.7* illustrates that the life cycle of many human activities begins slowly, experimentally and falteringly, moving through a period of rapid growth and expansion to a final slowing down and decline. Handy believes that managers need to know where their organizations are on the curve so they can start investing in a new curve during the middle phase of rapid growth, well before the decline sets in. Existing and new curves need to co-exist, the first generating the income needed to invest in the second. Both continue until the first curve enters its period of decline at which stage it should be replaced

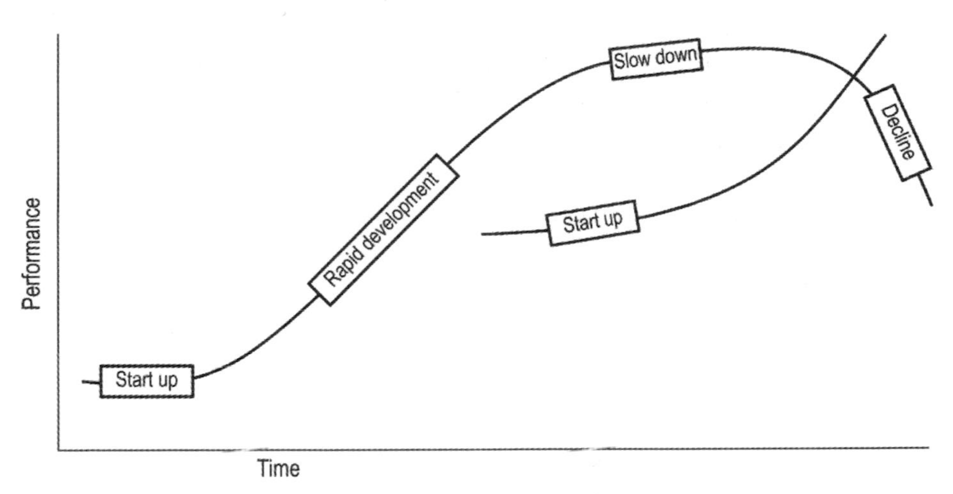

Figure 3.7: Life cycle of technologies

by the new one. Devising new curves requires existing assumptions to be questioned and leads to continuous experimentation and improvement, an essential ingredient of success.

These skills are not well developed in Japan, hence their current difficulties in moving away from previously highly successful technologies to new ones. Never-the-less, Kaizen has provided a powerful tool that has helped teams throughout Japan steadily improve their performance in many different situations. There is undoubtedly considerable scope for it to be used in many Western industries, including construction.

Partnering

The first practical evidence that confirmed the effectiveness of Japan's use of cooperation came from various studies of the international car industry. The most influential of these is described in Womack *et al.* (1990). This work provides a mass of detailed evidence about how and why the leading edge of the car industry moved from approaches based on US-style management to Japanese cooperation. Japan's approach develops long-term relationships which empower everyone involved to concentrate on adding value for the customer. This is achieved by applying Kaizen to every assembly activity and the supply chains that support them, to relentlessly eliminate waste wherever it is found. The resulting lean production now dominates car production throughout the world.

The cooperative methods on which lean production depends are called 'partnering' in the West. They first emerged in the car industry in the USA as a direct response to the greater success of Japan's methods of management. Their success in helping the US car industry fight back has led to partnering being adopted in other industries. It is an integral part of the management approach now used by leading manufacturing companies world-wide.

Partnering is influencing construction, in part because of its wide use by the industry's customers but also because research shows Japanese construction to be efficient. This is not because workers in the Japanese construction industry are better than those in the UK or the USA. The Japanese achieve better performance because their cooperative, long-term relationships empower workers at all levels to do their best work. In contrast, Western construction industries erect competitive barriers that take time to negotiate, reduce motivation and drive up costs.

The Japanese experience shows that given the right environment, construction can be managed in the same way as other modern industries. What is called partnering in the West is central to Japan's inherently efficient systems.

Partnering in the USA

On the face of it, major changes in attitude would be required before partnering could be accepted in the USA's highly individualistic, market-based form of capitalism. There are very few cooperative arrangements between industrial companies, largely because of tough anti-monopoly legislation. Major firms are owned by share-holders who trade their shares in a highly competitive stock market. All firms are at risk of take-over and so there is a high premium on short-term profits and dividend pay-outs to sustain share price. Firms' relationships with their suppliers are market-driven with contracts awarded to the lowest bidder.

The system appears to be almost totally pure capitalism. However, as Hutton (1995) describes, the USA has important shelters against the full blast of competition. At the federal level, many of the institutions put in place as part of Roosevelt's New Deal mitigate the competitive instincts of the US financial system. Also, federal spending on R&D is often used to support new high-tech industries, for example defence and aerospace. This comes close to providing an industrial policy. The Buy America Act ensures that public procurement is focused on local suppliers. At state level, banks are forced by restrictions on their activities to support their local customers and so they invest long-term to help them develop sustainable businesses. This is further reinforced by state governments who use procurement policies and subsidies to help develop local firms.

These various measures reflect a value system that formally celebrates individual rights, competition and the primacy of the markets but that is combined with a powerful impulse towards national solidarity which signals a more cooperative tradition in the US culture. The cooperative impulse was undoubtedly a factor in the rapid development of partnering as a response to competition from Japan.

There were, however, difficulties in applying partnering in the USA. As Kanter (1989) describes, partnering places new responsibilities on managers. They have first to help workers at all levels understand that individuality has had its day and success comes only when the team as a whole succeeds. They need to recognize that firms are increasingly interdependent and, whilst crude competition between independent workers or organizations provided benefits in the past, amongst today's interdependent firms it does not work. This is especially true where pressure from rapidly changing markets and new technologies makes it necessary to find new answers. Creativity requires a degree of security; it is killed off by the insecurity that results from crude competition inside any one organization or supply chain.

A crucial but essential change for many organizations is to share information openly with the workforce, customers and suppliers. The commitment of top management is essential if this is to happen because, at lower levels of work, people are naturally guarded and defensive of their own team's interests. So firms have to invest in teaching managers and workers to consider processes that flow through many teams, rather than allowing their thinking to be restricted by formal organization structures. Relationships and information that help achieve agreed objectives are what shape effective organizations. This forces many managers and workers to learn how to work through cooperative, face-to-face communication. Internal relationships often need to change in response to the realities of work and so senior managers have to give teams the flexibility to combine units in different ways.

External relationships also need to change. The old adversarial ways of dealing with outsiders have to be replaced by pooling resources and linking systems to exploit opportunities or solve problems. These are massive changes for organizations using paranoid styles of management fostered by fiercely competitive environments. Managers throughout the USA have had to learn that arms-length contracting, based on competition between many rival suppliers, may minimize price in the short term but only at the expense of cutting innovation and quality.

Many industries have recognized that cooperative relationships with a few carefully chosen suppliers provide the best approach. This approach provides more flexibility than direct ownership by enabling specialized resources and systems to be made available fast. It allows new ventures to be undertaken without all the costs associated with internal reorganizations. It also retains the vitality that comes from the individual ownership of small entrepreneurial organizations.

Kanter identified three forms of cooperative relationship in the USA which she calls service alliances, opportunist alliances and stakeholder alliances. Service alliances are consortia formed to provide a service too expensive for any one member to provide alone. Common examples include research, development, specialized insurance and coordinated demand. A common reason for consortia to be formed is to develop an expensive new technology to the prototype stage.

Opportunist alliances allow new ventures that either party would find difficult to undertake alone. A common pattern is for one party to contribute the technology and the other the market experience.

Stakeholder alliances are strategic partnering between the key actors in a value chain. They tend to focus on quality and innovation in creating joint systems. They seek continuous improvements, use joint planning, share technology and provide information openly. This kind of committed partnering involves substantial internal changes. The biggest change is in power. Other firms become welcome, equal partners, not manipulated adversaries. This alters internal power and requires managers to develop different skills and knowledge. It is no longer acceptable to complain about problems, now everyone is responsible for solving problems together. Purchasing, design, production and marketing have to work together more than ever before.

With partnering, managers face new kinds of issues. Open communication with a partner requires open communications inside the manager's own organization. In many large organizations this does not exist and can leave partnering managers looking foolish or devious when their promises are contradicted by the actions of another department. Managers spend more time outside the parent organization in meetings, exchanging data, learning about other organizations, and so on. As a result, internal support staff have a reduced role because much of their work previously involved dealing with outside organizations on the basis of arms-length contracts. These interactions are now dealt with by line managers who need to balance inside and outside interests in ways not previously necessary. Line managers become more competent and have a broader focus but there are fewer of them. These changes have been accelerated by information technology as routine jobs are done by computers and much of the administration associated with communication has become unnecessary as direct links between firms are set up. Many firms are unable to expand their markets fast enough to avoid savage downsizing and thousands of middle management jobs have disappeared.

The managers who remain now have to juggle constituencies rather than control subordinates. They can no longer make quick decisions, they have to consult and build consensus. Rather than giving orders, they need to discuss and search for agreements. Authority gives way to influence, command gives way to negotiation.

In a new partnering arrangement, managers' actions are scanned for clues about their real motives. Integrity is crucial. Managers need to create symbols of equality so that everyone feels part of a team. For these to be effective, they have to be based on an understanding of other peoples' frames of reference. So managers need social as well as technical and business skills. Effective managers do not evaluate, advise, interpret or probe, they seek to understand workers' feelings and work with them in devising new ways of working aimed at shared objectives.

Many firms faced with increasingly rapid change have invested in multi-skilling to create flexibility. Kanter reports cases where multi-skilling has helped to cut production costs by between 30 per cent and 50 per cent. In the best examples, firms have introduced profit-sharing schemes that reward workers for acquiring new skills. A number of approaches are used, including cash bonuses, stock options and profit sharing tied to audited profits accruing from workers' ideas. A common element of the most effective schemes is allocating rewards to teams and letting them decide how they should be shared amongst the individual team members. In most cases the allocation is equal but slackers do get weeded out.

Profit-sharing schemes have to be generous to encourage real innovation and, in effect, create new businesses inside an existing organization. This supports internal entrepreneurs who will expect to be paid as if they were owners developing a new business. This raises problems of ownership because it is doubtful if any new idea really belongs to just one person. This is a problem that construction firms need to address if they are not to face steady attrition with their most talented people leaving to set up new firms that become direct competitors.

Selling the idea of partnering is essential and difficult. People outside of a partnering arrangement often feel isolated and threatened. Much that used to be dealt with inside the formal management hierarchy is now handled by joint processes. Direct communication and decision making can cut out layers of internal communication, which is where many of the economies in partnering come from. Non-partnering colleagues often see the new arrangements as a sell-out to competitors or customers. It is important to keep them in the picture so they can understand the overall benefits. Cooperation should be rewarded, irrespective of which team contributes most. All must win on a basis that everyone can regard as fair.

Partnering in practice in US construction

Construction Industry Institute (1991) describes how partnering in construction began in the USA in the mid-1980s taking the form of strategic partnering by major private-sector customers. This was followed in 1987 by the Corps of Engineers applying partnering to individual public-sector projects. Traditional methods based on competitive tenders, tough contracts and formal administration were failing, litigation was a major problem and many projects were delivered late and over cost. The Corps decided that they needed to adopt a more cooperative approach in dealing with contractors and partnering was seen as the way to provide it.

A particular problem in introducing partnering in the USA was laws that require public bodies to select contractors for construction projects by means of competition, except in unusual or very urgent circumstances. Public bodies cannot combine projects to create a serial from similar projects. Indeed, if a contract can be divided, each subdivision must be the subject of separate competition. Despite this emphasis on competition, partnering is widely used on individual construction projects in the public sector. This is achieved by using competition to select contractors and then setting up partnering arrangements with the lowest bidder. Contractors are given the opportunity to refuse to partner but they rarely do, probably because, as Lancaster (1994) describes, case studies show that project partnering provides benefits for everyone involved and also because contractors are reluctant to risk offending powerful customers. An important benefit in the litigious climate of the USA is that partnering leads to substantial reductions in claims and litigation.

Bennett and Jayes (1995) describe cases where strategic partnering has been used in the USA; this is entirely in the private sector and it delivers substantial benefits, including cost reductions of up to 30 per cent, similar reductions in construction times, improved quality and greater certainty for customers and construction firms.

There are features of the US construction industry which made partnering relatively easy to introduce. US managers use management theory as a natural and vital part of their tool kit of ideas. This acceptance of and easy use of theory is one of the reasons why management ideas, including partnering, tend to be developed and applied readily in the USA. Also US managers take a very positive attitude towards problems. The common saying 'it's no problem' encapsulates an important feature of US culture. This is evident, for example, in managers' concentration on success and their emphasis on giving explicit rewards for team achievements. As a result there is a ready development of team spirit once a project team is assembled. The competitive attitudes needed to win the work are subsumed in a recognition that the only way any individual firm can succeed is if the project as a whole succeeds. It is at least arguable that partnering represents a relatively small extension of this normal team spirit and this is why it has been possible to apply it fairly quickly to construction projects in the USA.

Partnering is put into practice through the use of workshops. Partnering workshops are normally two-day meetings of all the parties involved in a partnering arrangement and so may deal with strategic issues or be confined to one specific project. Workshops usually take place at a neutral venue, with the help of a partnering facilitator. Workshops allow the team to get to know each other in a relaxed environment and help develop cooperative attitudes. Good facilitators recognize that teams develop by working together and so they move fairly quickly into the real issues of agreeing mutual objectives and deciding how they are going to work together, including how problems will be resolved. They set targets aimed at ensuring that their joint work produces measurable improvements in performance. In the main the measurements used are subjective, although Bennett and Jayes (1995) describe some examples of objective measurements of quality, cost and time.

The common use of workshops has given rise to a new profession in the USA, that of partnering facilitator. The facilitator's skills are similar to those used by facilitators in formal negotiations, value management workshops and team-building exercises. Partnering facilitators need, in addition to general facilitator skills, an understanding of construction and of the concepts of partnering. In the USA, most partnering workshops use an independent facilitator and it seems likely that the success of project partnering depends on that person's involvement. Workshops give rise to some initial costs but the US experience is that these are offset by the benefits many times over.

Bennett and Jayes (1995) use case studies of US practice to identify three features critical to the success of partnering. The first is that partnering teams agree mutual objectives. These should be designed to give everyone associated with the project a realistic expectation that their work will be successful in business terms. This recognizes that firms make more money by working together towards agreed mutual objectives than if they pursue individual interests in a competitive, adversarial manner. In other words, it is more efficient to concentrate on getting the work done than it is to argue over who is to blame for problems.

Obviously in such a complex activity as designing and constructing a new building, problems will arise. So partnering accepts that, even with agreed mutual objectives, problems will have to be dealt with. So the second essential feature of partnering identified by Bennett and Jayes is to agree how the problems that inevitably arise will be resolved. The aim in this second feature of partnering is to deal with problems quickly, at the point at which they have arisen and so prevent them from festering and growing into disputes. This requires a cooperative search for solutions that leave everyone feeling their interests have been properly taken into account. It must not be based on one party imposing an answer on the others. It must not be based on legalistic arguments about rights, duties and liabilities. There should be no place for trying to blame individuals for problems that have their root causes in the total system. In other words, in solving problems partnering teams accept joint responsibility for their projects and their outcomes.

The third essential feature of partnering is that the partnering team actively seek to achieve continuous improvements in their performance. The improvements should be measured and should provide better value for money for the industry's customers and higher profits for everyone in the partnering team. In many cases total quality management is used to provide the techniques needed to achieve continuous improvement. Its focus on analysing and improving processes has a similar emphasis to Kaizen.

Bennett and Jayes (1995) take the view that these three features, mutual objectives, cooperative problem resolution and continuous improvement, are essential to partnering. The three features enabled the US construction industry to balance its traditional competitive approach with more cooperative methods and, by so doing, it had achieved significant improvements in performance by 1995.

Partnering in the UK

The use of partnering in the USA influenced thinking in UK construction. The UK construction industry faces a political, economic culture that has most of the competitive features of the US approach without the compensating impulse towards an inclusive national solidarity. Hutton (1995) describes the resulting gross inequalities in income, increasingly insecure employment conditions for the majority and financial institutions narrowly focused on the highest levels of short-term profit. These weaknesses are reinforced by highly centralized institutions run on the basis of primitive forms of democracy in the interests of a self-serving elite. Private interests and the market are celebrated as the only efficient and responsive forms of organization, while notions of cooperation, common interests or public spirit are dismissed as bureaucratic, interventionist or socialist.

As a result there are deeply rooted structural reasons that lead the UK construction industry to concentrate on flexibility, short-termism and adversarial attitudes. Thus, the UK provides a uniquely inhospitable home for partnering but, equally, is more in need of the changes it provides. This case is well stated in Latham (1994).

The need for change was fully reflected in the New Labour Party's overwhelming victory in the 1997 General Election on the basis of a radical manifesto. As far as construction is concerned, one of the new government's first actions was to set up a Construction Task Force chaired by Sir John Egan to report on the scope for improving the quality and efficiency of the UK construction industry. The result is Construction Task Force (1998) which set the challenging targets for improvement listed in *Table 1.3* and which recommends the industry should rethink construction or, in other words, should adopt a new paradigm based on using lean production and partnering.

The emergence of partnering in UK construction

The emergence of partnering in UK construction was researched to provide background information for Construction Industry Board Working Group 12 (1997). The following descriptions draw on that unpublished background work undertaken by the author with two research assistants, Ian Ingram and Sarah Jayes, during 1996.

Partnering was first applied in UK construction in the North Sea oil and gas industries. Change was forced on these industries in the early 1990s by a combination of low market prices and increasingly high operating costs. These powerful market forces meant that the industries needed to move away from a culture in which the primary aim was getting oil and gas ashore almost irrespective of cost, to one in which efficiency was essential for survival in world markets.

The UK oil and gas industries faced very real barriers to the introduction of partnering. These included the industry's traditional focus on the short-term self-interests of individual firms, and a continuing attachment to the blame culture implicit in adversarial customer:contractor relationships.

The primary concerns of the major firms in the industry, short-term profits and dividend payments, worked against the development of long-term relationships. Internal politics and conflicts, including departmental and group rivalry, made ideas of cooperation seem ridiculous. It was common for project teams to be made up mainly of short-contract hired personnel who had no personal interest in long-term, cooperative relationships. There was a widely held view that performance improvements would benefit only shareholders and directors at the expense of workers. Finally, the management systems commonly used in the industry did not provide measured feedback on the factors critical to success.

The cultural changes required to overcome these barriers have been significantly difficult to achieve and probably have succeeded only because there was massive scope for improvement as demonstrated by, for example, Shell who reduced construction costs by some 34 per cent over a four-year period in which the profit margins of the partnering contractors increased from 6 per cent to about 11 per cent. Also, British Gas have partnered with the main design consultants and contractors on very large individual projects and achieved construction costs 20 per cent lower than their best ever previous performance.

Case studies of partnering in the UK oil and gas industries suggest that the initial efforts in partnering need to concentrate on building a common culture and understanding, and improving total business processes through the development of mutual objectives. It is normal for these cooperative activities to take place at partnering workshops guided by experienced facilitators. Carlisle and Parker (1989) describe how the most successful of the facilitators have developed unusually sophisticated methods of encouraging cooperation in order to overcome the short-term, adversarial attitudes that are so prevalent in the UK. Their own approach includes a game called the red:blue game, based on the Prisoner's Dilemma. It helps teams understand the implications of choosing to act competitively or cooperatively.

The case studies also suggest that, in the UK, mutual objectives are much more likely to be effective if they are reinforced by profit-sharing schemes which allow everyone to earn bigger profits when projects go well. As a result, sophisticated profit-sharing schemes have been devised and are widely regarded as being a significant factor in the improvements achieved in the North Sea oil and gas industries. This heavy reliance on profit motivation is a clear indication of the difficulties of introducing partnering in the UK. There are examples of more modest, yet effective, incentive schemes and some consultants and contractors feel that an assurance of regular workload alone is sufficient incentive. Never-the-less the most significant benefits have been in arrangements based on carefully designed profit-sharing schemes.

The civil engineering sector of the construction industry in the UK should provide a relatively easy context in which to introduce partnering because consultants and contractors have similar engineering backgrounds in terms of qualifications and training. This should help ensure a commonality of language, and mutual recognition of each other's requirements. Bennett, Ingram and Jayes' unpublished case studies undertaken during 1996 indicated that considerable benefits were being achieved, including reductions in project times of up to 20 per cent and a wide range of cost reductions of up to 25 per cent compared with traditional approaches.

The UK building industry provides a challenging context for partnering because of the way it is fragmented into a multitude of different professions. This causes misunderstanding, disagreement and very little commonality of purpose. Early attempts at applying partnering in the building industry showed that adversarial attitudes formed a large barrier. Case studies suggest that cooperative attitudes can be developed in the UK but several workshops are necessary before traditional attitudes are relaxed, and people begin to trust each other and be reasonably open about their interests and concerns.

Partnering in the building industry highlights the need for firms to make internal changes so decisions to adopt cooperative relationships with other firms are not undermined before they develop. Bennett and Jayes (1995) describe this as internal partnering, which they found to be vitally important. They report cases where internal partnering was neglected resulting in partnering arrangements that had begun to provide improvements in performance being abandoned by senior managers who still believed in the superiority of market-based competition.

Bennett and Jayes (1995) draw on case studies of long-term relationships in the UK building industry, only a few of which had been formed into formal partnering arrangements. They found the largest benefits in these informal approaches occurred in design and management processes, rather than in the direct construction work. This suggests that there is considerable scope for improvement in work which depends on exchanging information, so that as people work together they become more efficient at understanding each other and can achieve improvements in efficiency. This happens even in informal long-term relationships. However, Bennett and Jayes found that improvements in direct construction activities, which account for the largest part of total costs, are more difficult to achieve and need the full disciplines of partnering.

Early attempts to use partnering in the UK construction industry threw up some remarkable misconceptions about partnering. Most worrying were cases where customers, who have traditionally used ruthless competition to push prices down to levels that provide absolutely minimal profit levels for contractors, were trying to use partnering as a way of forcing even more out of the industry at no risk to themselves. These are classic examples of win:lose situations where the customer wins and is not bothered that consultants and contractors lose. This is the antithesis of partnering. It results from deeply ingrained habits of secrecy, an unwillingness to trust other people, a narrow concentration on individual interests and an instinctive reliance on competition. All of these primitive, competitive habits are prevalent in the traditional approaches of the UK construction industry and they work against partnering.

As a result, early attempts to introduce partnering in the UK construction industry required careful preparation. This usually concentrated on finding facilitators to guide partnering workshops, devising ways to measure the success of projects and developing profit-sharing schemes to reinforce mutual objectives. These initiatives provided some benefits but did not directly address the main purpose of partnering, which is to radically improve the performance of the industry.

The early use of partnering made it apparent that most managers in the UK construction industry did not have a working knowledge of the techniques needed to analyse their processes, were reluctant to exchange information in a manner that would support benchmarking, had no objective measures of their own performance and only occasionally used creative techniques such as value management. Thus, they had no effective ways of making the changes in performance necessary for the industry to meet the legitimate demands of its customers. Inevitably, Bennett and Jayes' early case studies provided very few examples of continuous improvement because the construction industry had no systematic basis for producing it.

Bennett and Jayes (1995) therefore describe an elementary approach to partnering. Never-the-less it has been widely used in the UK construction industry. The report is based on research into Japanese construction and case studies of partnering in US construction. The UK view came from case studies of long-term relationships between customers and construction firms. These included relationships using serial contracts in the public sector and long-term alliances in the private sector. The report remains a good introduction to using partnering. It describes partnering as based on mutual objectives, an agreed method of problem resolution and an active search for continuous measurable improvements. Subsequent research shows that problem resolution is too narrow and too negative a concept and, in their subsequent report (Bennett and Jayes, 1998), the second essential element of partnering is changed to agreeing the decision making system.

Developed partnering practice in the UK

The report by Bennett and Jayes (1998) is based on some 200 case studies of partnering in the UK construction industry. It defines partnering as a set of strategic actions which embody the mutual objectives of a number of firms. These are achieved by cooperative decision making aimed at using feedback to continuously improve joint performance.

This recent research shows that partnering can be applied in the UK construction industry and is significantly more efficient than traditional competitive methods. The benefits are not achieved overnight; partnering has to be built up step-by-step over many years. Bennett and Jayes (1998) describe three distinct stages in the process. The first stage is formed by the construction businesses and their customers who have used the model of partnering described in Bennett and Jayes (1995). These cases show that this early report described what is essentially a project-based approach, even when it is applied to a series of projects. Bennett and Jayes now call it 'first generation partnering'.

Their second generation partnering is well established amongst leading firms. It consists of partnering by a group of consultants and contractors who add a long-term strategic dimension to a series of projects for one customer.

Bennett and Jayes found a further development in partnering which they regard as forming a third generation of partnering. This is spearheaded by construction firms organizing their business to provide continuity in their workloads. They are doing this by using partnering throughout their supply chains to produce products designed for specific categories of customers. These are marketed so as to provide a steady stream of work. Most of the cases of third generation partnering found by Bennett and Jayes were based on the UK Government's private finance initiative. This provides arrangements by which private-sector firms undertake new development work and lease the resulting facilities to public-sector bodies. Hospitals, prisons, student accommodation at universities and other public-sector facilities are being operated by private-sector firms in this way for periods of up to thirty years.

Bennett and Jayes (1998) describe research which shows that leading practice in each of the three generations delivers progressively higher levels of cost and time benefits. The main results are given in *Table 1.2*. They suggest that when project teams work together within a cooperative partnering arrangement, they are able to be more efficient than with traditional, competitive methods where everyone is expected to look after their own interests and hope that the result will be a good project. To use Carlisle and Parker's (1989) metaphor, firms are better off when they work to make the pie bigger than when they fight to get a bigger share of the existing pie. This is true as long as they make sure that everyone gets a fair share of the bigger pie. Making the pie bigger means cutting out waste, searching for savings, finding better ways of working, creating better designs, finding new technologies and devising more efficient processes. In other words, delivering more for less. Giving everyone a fair share means everyone gets a fair profit, covers their fixed overheads and all the other costs involved in doing their best possible work for the project. And yet, because partnering is more efficient, the customer pays a lower price.

This suggests that people either concentrate on competitively looking after their own interests because they feel they will be cheated if they do not, in which case they are not able to work efficiently; or they feel their interests are safe and so they can cooperate in doing their best work for the good of the whole team. This can lead to massive improvements in efficiency, profitability and customer satisfaction. The following chapters put these practical issues into the framework of the new paradigm which provides the best currently available basis for understanding how the cooperative behaviours encapsulated in partnering can be balanced with competition to deliver long-term benefits to customers, consultants and contractors.

New framework

4

Patterns, structures and processes

The new paradigm means that teams responsible for construction undertake their work through networks. Capra (1996) identifies the key characteristics of networks as patterns, structures and processes. Patterns are model configurations of relationships that define organizations and shape the way they can work. Patterns are given practical expression in an organization's structures and processes. Any given pattern may be expressed in structures and processes that differ considerably in the practical detail of real-world organizations. This chapter describes the patterns teams should keep in mind in making decisions about their organizations. The issues that influence decisions about structures and processes are described in Chapters 5 and 6.

Patterns are internal sets of ideas which help us make sense of the external world and our relationships with it. The more that the patterns used by teams are in tune with reality, the more successful they will be. Humans learn about reality in many ways. Science, art and religion all provide aspects of this knowledge using very different methods. The construction industry has to work with whatever is currently the best understanding of reality. That is often a contentious issue; people disagree about science, art and religion and new knowledge often comes out of those disagreements. People working in construction have to make judgements about which views they accept and which they think are wrong or unhelpful.

This chapter contributes to these decisions by describing this author's views about the patterns of relationships that should guide the design of the structures and processes used by construction organizations. It provides a framework of ideas that research suggests will help teams responsible for construction to work effectively. However, people have to work out their own patterns, their own way of seeing and understanding the world. Written descriptions, like this one, help but direct experience and thinking about what it means provide the most powerful patterns.

New patterns of practice

The new paradigm suggests that the construction industry should be seen as a tapestry of richly interconnected networks made up of human and physical resources and the relationships between them. The networks that form the industry interact with their environments which are also formed of richly interconnected networks.

Given competent people supported by appropriate physical facilities, the industry's performance is determined by the patterns of relationships between them, and between them and their environments. Individual success in isolation from the networks of which we form part makes no sustainable sense. It is ephemeral, an illusion; lasting success can come only from the success of the communities of which we are part.

This is an important point because individualism provides powerful motivations. But success at the expense of the wider community is hollow. It has to be defended by security systems, police and laws directed by the interests of private property rather than by justice. It brings only temporary happiness. Real happiness comes from being part of a community in tune with nature, including human nature's highest aspirations. Maslow (1954) famously described the human hierarchy of needs, which is summarized in *Figure 4.1*. It tells us that only lower-level needs are satisfied by possessions. They merely provide a platform for individuals to develop and realize their own potential. The height of the lower levels of the hierarchy of needs creates a threshold that determines how soon we can begin to concentrate on personal development. Greed is damaging, because it raises the threshold too high and people cannot spare the time from the task of accumulating more possessions to seek their own fulfilment. Cooperating in communities that take account of everyone's interests is the most effective basis for individuals to progress to the higher levels of Maslow's hierarchy of needs and so begin to satisfy their own highest aspirations.

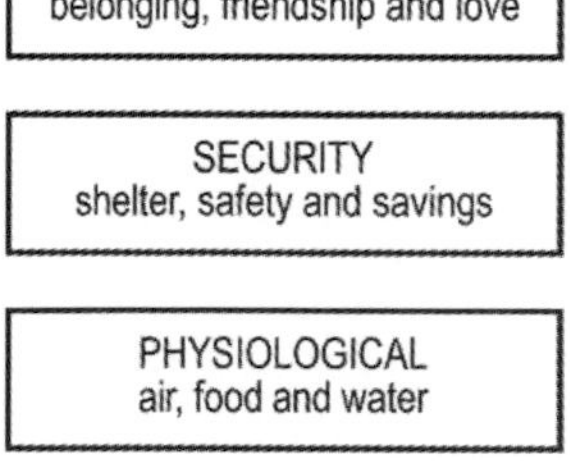

SELF-ACTUALIZATION
fulfilment of life's purposes

ESTEEM
pride, respect and status

SOCIAL
belonging, friendship and love

SECURITY
shelter, safety and savings

PHYSIOLOGICAL
air, food and water

Figure 4.1: Maslow's hierarchy of needs

However, competition is inevitable because the larger network of which construction forms a part seeks its own survival. In doing so it may damage the construction industry, or parts of it, in the wider public interest. Threatened subsystems seek their own survival and so competition arises. Communities then have to decide which interests will prevail or find win:win solutions. Change usually produces some winners and some losers. It is in individuals' best interests to accept that communities should protect those who would otherwise lose out because there is no way of knowing who the losers will be when the next change occurs.

An important implication of there being winners and losers is that organizations have to decide the extent of the community in whose interests decisions are to be made. It should include everyone who reasonably believes they are affected by the organization's work or decisions. Practice is moving towards this view as organizations everywhere increasingly try to involve customers, suppliers and local communities in their decisions. This recognizes the reality of how organizations are formed in a networked world.

The construction industry comprises rich patterns of relationships, some of which are reinforced and supported because they recur and are found to be helpful. Some links give rise to boundaries that limit the formation of new relationships. The nature of the links near their boundaries make organizations more or less open to new relationships. Strong links and boundaries give rise to organizations at all levels of work. The basic organization unit is teams of individuals who between them provide all the skills and knowledge needed to undertake specific types of construction work. The reasons for this are described in Chapter 5, as is the diversity of teams that form modern construction industries.

Specialized teams interact to form construction organizations, including those responsible for projects, supply chains and long-term development. Overlaying this basic structure are firms, specialist disciplines, professions, associations, institutes, clubs and many other more-or-less permanent sets of relationships between individuals. They play a helpful role in so far as they work in support of teams that undertake direct construction work.

Feedback is crucial to the survival of all these emergent organizations. This is a key idea in the new paradigm; if parts of the networks have effective feedback loops, they form systems capable of taking actions directed towards their own long-term

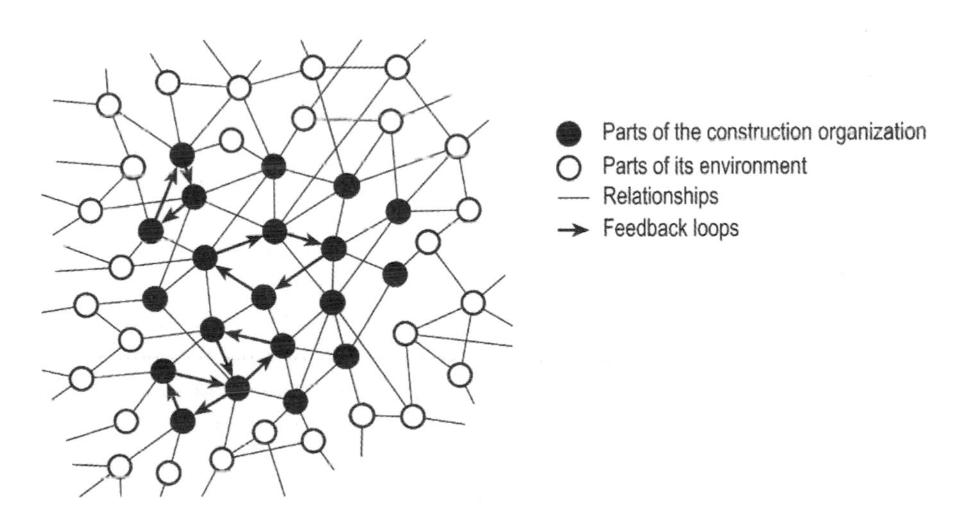

Figure 4.2: The basic pattern of construction organizations

sustainability. The behaviour of these self-organizing systems depends on patterns of internal and external interactions. This means the behaviour of construction teams and organizations can be understood only in the context of the wider industry. In the same way the industry can be understood only in the context of the wider communities of which it forms part. *Figure 4.2* illustrates these basic patterns.

The first important practical consequence of the new paradigm is that everyone involved in the industry should take the self-organizing capacity of systems into account in making decisions. They will find it easier and more efficient to work with the natural instincts and habits of systems than to change them. This means, for example, that relationships which the people involved find helpful in carrying out their work should be encouraged and supported.

The new paradigm sees communication as the way in which behaviour in teams and organizations can be coordinated and directed towards agreed objectives. This means, because of the self-organizing capacity of systems, organizations can most effectively manage their work by wide-ranging communication about their values and aims. People at all levels should ask questions, listen, debate and seek to build consensus. Shaping their organization's perception of its own best interests in this way aligns its self-organizing capacities with the agreed values and aims. Decisions should be based on wide discussion by all those affected by the outcome. In these discussions it is important to encourage and respect diverse views. There are usually several valid ideas about the best way forward. As long as they do not infringe anyone's basic human rights to happiness and freedom from suffering, they all deserve to be considered. In the long run it is more efficient for decisions to be widely debated to build consensus than to have a manager make a narrowly based decision only for it to be challenged and undermined as soon as it is announced.

Communication is crucial to decision making. Open information used by competent people trying to achieve the best for themselves and all the organizations and communities they influence is the most effective basis for achieving the greatest public good. Secrecy and narrow self-interest are nearly always damaging for everyone involved.

The next main implication of the new paradigm is that cooperation should be seen as the guiding force in all interactions. It, too, depends on open communication, which is most likely in well established relationships. This is because it takes time to learn how to communicate effectively with another person, and it takes time to learn how far they can be trusted with open information. Which is why long-term relationships are efficient and should be used wherever possible. Where they do not exist, extra attention should be given to communication. For example, occasional customers need clear, objective information about the choices provided by the construction industry. New project teams should give extra attention to agreeing how they intend to work together and how problems will be resolved.

As Chapter 2 explains, communication provides information about the performance of others in the industry which is an effective competitive spur towards improved performance. Indeed, this kind of well informed competition, blended with cooperation, is the most effective basis for joint action.

The general principle running through these immediate implications of the new paradigm is that actions should be based on discussion and consensus amongst those who make up the communities affected by the actions. This is real democracy, which takes time and care. All key decisions about the industry's work should be made this way. Since democratic decision making is slow and expensive, organizations should aim at finding robust answers to problems which can be used consistently until time and resources are available to find even better, carefully considered answers.

It follows that the industry should use established answers wherever possible. The costs involved in finding new answers should be incurred only when none of the industry's established standards or procedures are acceptable. Also as far as possible, new answers should provide measurable improvements to established ways of working, not merely a different approach. There are occasions when a new answer is needed but these should be infrequent and the costs of abandoning established answers should be accepted only after careful debate and discussion.

Accepting these views means that the industry's work is seen in terms of mainstreams and new-streams that are linked, as *Figure 4.3* suggests. Mainstream projects are sufficiently predictable to justify aiming at controlled efficiency using standardized answers. New-stream projects should encourage creativity and innovation in dealing with unusual demands from customers, researching potential opportunities and exploring new ideas. The results should contribute to improvements in mainstream performance. It is important also that in total, new-stream projects provide a stock of new ideas and methods for the industry to draw on when it needs to deal with changes in markets and technologies.

Thus, the new paradigm provides a basis for the construction industry to offer customers clear choices. These will include standardized facilities and supporting services delivered quickly, reliably and fully complete at stated prices that provide competitive good value; and individual facilities and services produced on commercial bases tailored to customers' individual needs.

This new pattern of work serves to identify the key issues that should be

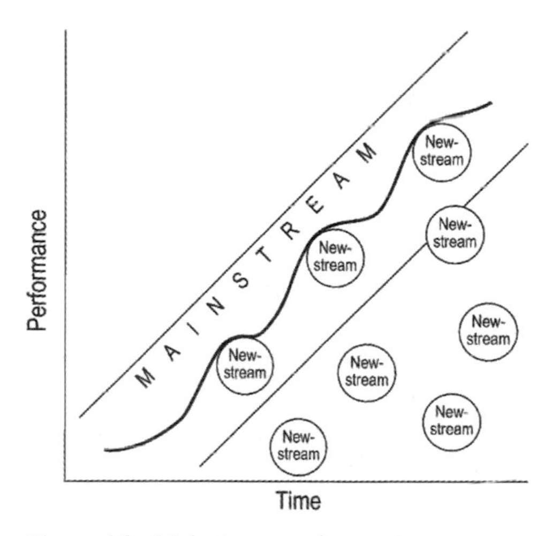

Figure 4.3: Mainstream and new-stream

considered in making decisions about the management of construction. The seven pillars of partnering described in Bennett and Jayes (1998) provide a useful framework for describing these issues. Each pillar provides guidance on one aspect of the patterns that should shape the work of organizations. The seven pillars are strategy, membership, equity, integration, project processes, benchmarks and feedback. Together they provide a framework, shown in *Figure 4.4*, that balances the strengths of competition and cooperation.

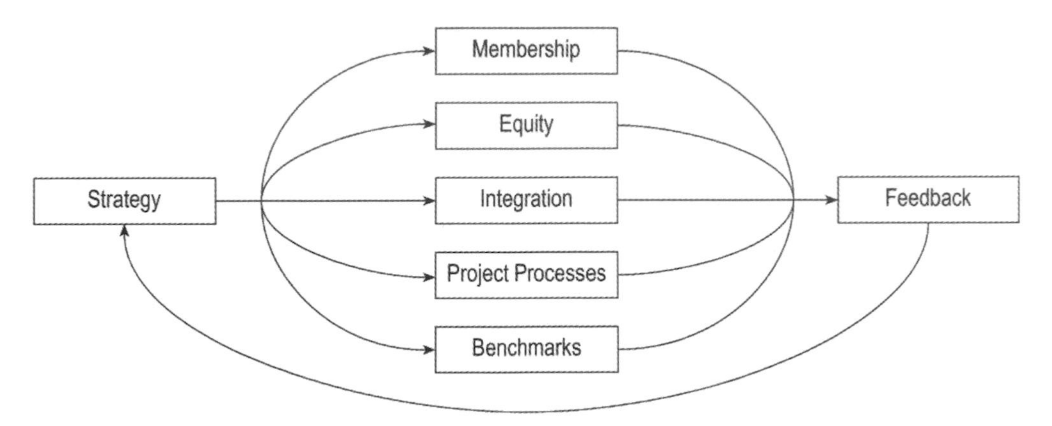

Figure 4.4: The seven pillars of partnering

The practical issues involved in designing effective structures and processes are described in Chapters 5 and 6. Never-the-less, it is helpful to assume a context for the idealized pattern described in this chapter. The following description assumes that construction products and services are provided for customers by integrated organizations. In general these organizations are formed by a number of firms working together using strategic partnering. In theory, the organization responsible for construction could be a major firm providing comprehensive construction services. However, in practice, even the largest of construction firms rely on consultants and specialist contractors to undertake some of the work involved in their projects and so it is assumed that the organizations responsible for construction are formed by a network of firms. It is assumed they use strategic partnering because it is the most effective approach to managing joint work by several firms. It is further assumed that each such organization has a general strategic plan established by a strategic team comprising representatives of the key firms in the overall organization. The general strategic plan is elaborated into objectives and targets that guide the work of task forces and project teams. The whole organization forms an intelligent, controlled system guided by many feedback loops.

Figure 4.5 illustrates some of the key features of such organizations. The most significant point is that even in the unrealistically small organization illustrated,

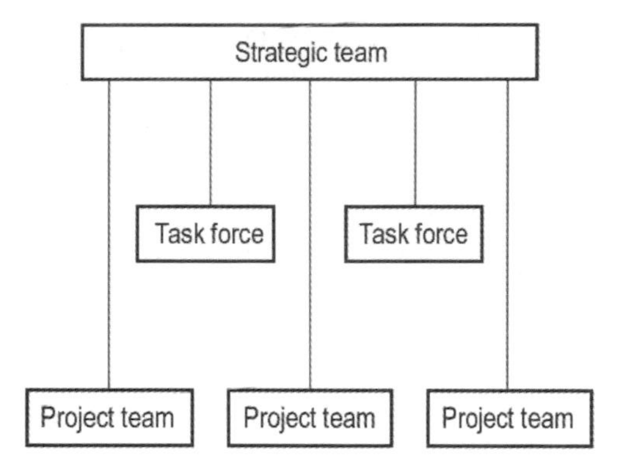

(a) Traditional management structure

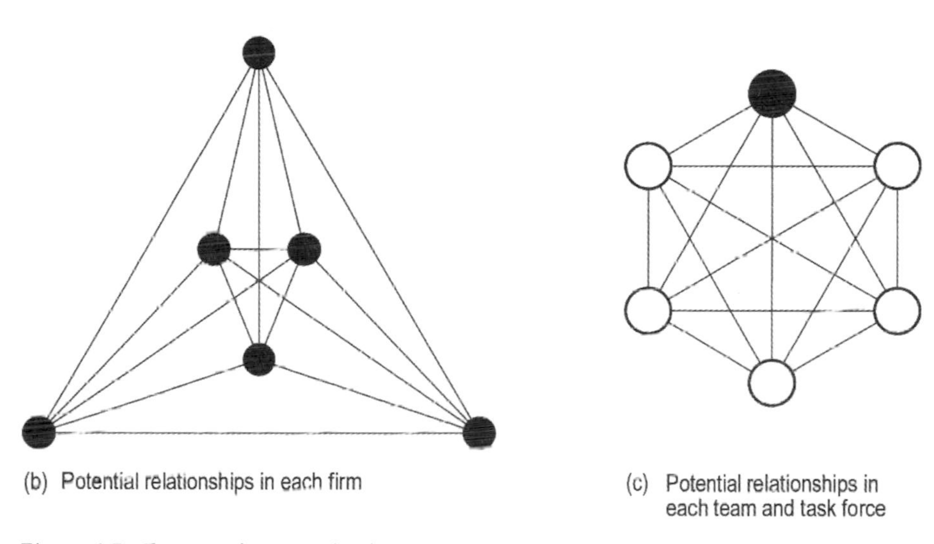

(b) Potential relationships in each firm

(c) Potential relationships in each team and task force

Figure 4.5: Construction organizations

the number of potential relationships is enormous. *Figures 4.5b* and *4.5c* indicate this complexity. With just six member firms undertaking three projects and setting up two task forces there are 630 potential relationships. *Figure 4.5a* shows the traditional management diagram that recognizes only five of these. In practice, many of the potential relationships will prove to be useful and so grow into strong links. There is no way to predict which these will be, so managers have no rational choice but to allow, and indeed to encourage, their organizations to develop organically. The following description of the seven pillars of partnering provides a framework for thinking about the key issues raised when construction is viewed in this way.

Strategy

The strategy pillar stems from the belief that firms forming a construction organization are more likely to be successful if they have a clear strategy. It does not provide any guarantee of success, but firms are more likely to achieve what they need from a relationship if they first agree an overall strategy which provides the key features listed in *Figure 4.6*.

STRATEGY
Long-term objectives
Medium-term targets
Overall structure
Main processes
Value system
Ethical framework
Common culture
Freedom to innovate

Figure 4.6: Key features of strategy

Strategies are formally produced and agreed by strategic teams. However, they should be based on wide-ranging discussions and consensus building throughout the organization and the communities of which it forms a part. In the early days of any construction organization the overall vision is shaped by one or two enterprising individuals but its development over time should take account of much wider interests, experiences and points of view.

Strategic teams should think long term so that strategies provide a long-term context for the targets set for project teams and task forces.

Strategies define the most important long-term objectives and establish medium-term targets directed towards their achievement. They establish the organization's overall structure and the main processes to be worked through by all parts of the organization. These are commonly described in the organization's policies, procedures and standards. They should not be seen as fixed and permanent because strategies evolve as markets and technologies change, and managers adopt new ideas. So the strategy should be reviewed regularly to ensure it continues to reflect everyone's best interests.

The strategy pillar needs careful attention because, generally, there is a lack of explicit strategic thinking in the construction industry. Strategies exist but they are implicit, emerging from operational decisions rather than being based on a clear sense of direction. This is largely because the way the industry's work is currently organized concentrates peoples' attention on individual projects. This needs to change because the development of long-term strategies is fundamental to creating successful construction organizations.

One of the key roles of the strategic team is to develop a common understanding of the organization's value system. This should help everyone involved understand the products and services that the organization aims to provide for its customers, and the terms and conditions on which it is prepared to do business. This defines the organization's ethical framework and provides the environment for a consistent culture to develop.

A key issue for the strategic team is to decide how they can provide the continuity needed to keep efficient teams together. The way to provide continuity is not to look for similarities in the end products but rather to devise common processes. This is how modern manufacturing firms are able to combine volume production with variety in their end products. They develop standardized processes operated by teams that stay together, and then design products to fit one of the processes. Products that do not fit are either not made or subcontracted to smaller bespoke manufacturers. This allows firms to shape demand to provide the continuity needed to invest long term in their structures and processes. *Figure 4.7* puts this approach into the context of a mainstream organization. The organization has a range of mainstream processes that meet the needs of broad categories of customers and accepts that some customers' needs are best met by a new-stream process which may be provided inside or outside the organization.

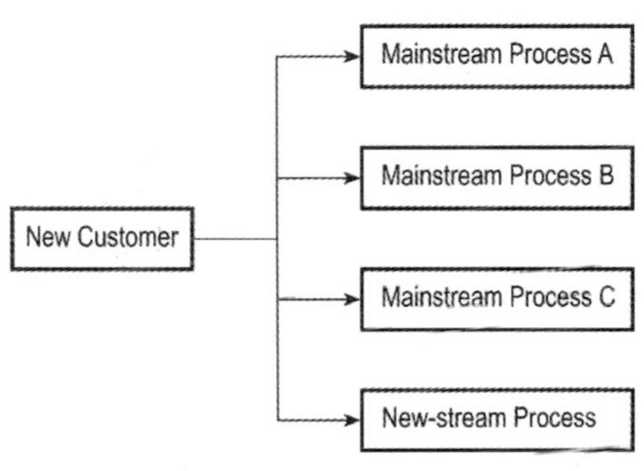

Figure 4.7: Strategy for dealing with a customer's needs

Strategies need to ensure that the organization becomes more and more competent at its chosen specialisms. This means becoming more and more efficient in its mainstream work. It also means ensuring that the organization has available the skills and knowledge necessary to deal with a wide range of problems as they emerge. The big Japanese construction companies all have large research institutes that undertake this role well. In Japan, when a project faces a difficulty the project team can call on an expert from the research institute to help find an answer. Strategies should also ensure, for the organization's long-term survival, that there is a pool of new ideas ready to provide the basis for a step change when one technology has run its course or the market changes fundamentally.

It follows that good strategies are not rigid; they cause managers to stick to agreed objectives, yet allow them a degree of freedom to act on their own initiative and provide that element of surprise which characterizes world-class creativity and innovation. This means living with ambiguities and sometimes accepting that rules can be broken by people dealing with specific situations or opportunities as long as they act in the best interests of the overall organization.

Project strategy

In setting targets for individual projects, strategic teams need to take account of the nature of the work being undertaken and the aims and perspectives of the firms involved. Then individual project teams can take these well considered targets into account in developing objectives that contribute to the organization's strategy as well as taking account of the individual customer's needs. In developing project objectives, it is good practice to aim at delivering more than the customer expects. This is possible only when the team clearly understands the customer's business case for the project, which should include a statement of the overall budget. This is not to argue that finance should dominate decision making. Indeed, many customers do not want a simple focus on the lowest price, they want the best value in their own terms. Never-the-less the budget provides an early indication of the value a customer places on the project's outcomes.

When members of the project team and the customer have worked together before, they are likely to have an implicit understanding of how their separate interests can be balanced. But when the customer and project team are building new relationships, particular effort should be devoted to ensuring a good understanding of each other's interests.

The first step for the construction organization in understanding a customer's interests is to identify exactly who the customer is. It is important to look beyond the customer's formal organization and identify all the internal and external stakeholders. For example, the need to satisfy funding institutions may limit effective partnering by imposing constrains on the technical answers and processes that can be used. Another important group whose views need to be taken into account are the users of the new facility. Particularly with a new building, new users often need help in understanding how it could contribute to their work. This is why experienced project teams bring in experts to help users understand the potential of new facilities. Neighbours and environmental and other pressure groups should be brought into the discussions. The interests of the latter two are, to some extent, reflected in the policies of regulatory authorities but often go beyond the formal legal framework.

Having answered the question 'Who is the customer?', the project team should consider what each of the people involved could get from the project that is exciting, surprising and goes beyond their expectations. All the interests should be considered in agreeing the project's objectives. This careful definition of objectives inevitably takes time and resources but, when a team bases the project objectives on an inadequate view of the customer's interests or of their own organization's strategy, all kinds of problems arise. For example, it is important to be clear whether quality standards can be altered in order to stick to the programme or budget. It is good practice to discuss these kind of trade-offs at an initial workshop before problems have arisen and agree the priorities that represent best value for everyone involved. Then everyone should stick to the priorities agreed at the outset, unless a further workshop has agreed to alter the objectives. Maintaining a consistent focus is often difficult in the heat of dealing with problems in a busy design office or on a complex site but without this consistency, uncertainty, cynicism and mistrust begin to grow. Project teams may well include construction professionals who believe their best interests lie in maintaining old ways of doing things. They will seize on any inconsistencies to justify resisting new ways of working. So managers need to be tough in establishing project objectives that take account of everyone's interests. Then they should insist that everyone sticks to them.

Project teams need to know whether they are likely to face changes to the project's objectives. Change may be necessary because the customer operates in a volatile business environment where new developments in technology, changing markets or new regulations force a fundamental rethink. Whatever the cause, the possibility of fundamental change in the customer's interests must be taken into account in the structures and procedures established for project teams.

A project's objectives should be expressed in an agreed mission statement and brief for the project. These should describe the problem or opportunity underlying the project. They should set out the carefully thought-out requirements to be satisfied by the end product, including its planned life and the support services provided to help users make full use of it. The mission statement and brief establish the constraints within which the project team is to work, including the quality, time and cost requirements. It is usually sensible for these key targets to be fixed overall but to allow flexibility in detail so the project team can search for the best possible answers. The mission statement and brief should also identify problems and risks that need to be taken into account by the project team. Finally, they define the success factors that will be used to measure the project's outcomes.

In developing agreed mission statements and briefs, project teams should take account of their own organization's strategy. This is guided by their understanding of the long-term objectives and current targets, and by the setting of targets for individual projects by the strategic team. In most cases these will require the project team to regard the project as part of the organization's mainstream work and so work through well established structures and processes in producing predetermined products. In addition they may be asked to put into practice one, or at the most two, specific improvements to performance that have been developed and tested by task forces. Some projects will be seen as new-stream but this should not be common because new answers are inevitably expensive, time consuming and risky. Never-the-less a proportion of an organization's projects should be set challenging objectives that require radical improvements to specific aspects of the organization's work. The decision to treat any given project as mainstream or new-stream should be taken by the strategic team only when they have a clear understanding of the customer's own broad objectives for the project. When these are consistent with the organization's strategy, a new-stream project can be authorized to develop new answers.

There are, of course, construction organizations who's strategy is to work only on new-stream projects. Once they are reasonably established, the mainstream of their work will use well developed approaches to project structures and procedures that provide great flexibility. The objectives for individual projects will emerge gradually as the work progresses and decisions are agreed. Even in this context there is scope for a strategy that deals with the long-term development of the organization's competencies. This may include investing in sophisticated information technology systems, refining creative techniques, devising flexible control systems, researching radical new technologies and other long-term developments. Without this, new-stream organizations rarely survive for very long. It is important for construction's future health that many new-stream organizations do flourish. They contribute new ideas that help ensure continuous improvements in mainstream work and provide the basis for step changes when they are needed.

Membership

The membership pillar is concerned with decisions about the firms that make up construction organizations. The firms should, between them, be capable of doing all the work likely to make a significant difference to their joint performance. In practice this usually requires many firms, each providing some key skills and knowledge. The main features of this pillar are listed in *Figure 4.8*.

MEMBERSHIP
Competent skills and knowledge
Strategic criteria
Clusters of firms
Open cooperative behaviour
Constructive competition
Continuity of profitable work
Higher profits
Lower prices
Benchmarked value for customers
Regular performance reviews

Figure 4.8: Key features of membership

The choice of firms to make up such organizations should be influenced by the long-term strategy and by relationships that form and are found to be productive. In this way there is both an overall sense of direction and an organic quality to the way the membership of construction organizations develops.

The strategy provides the criteria for deciding which firms become members of the organization. The criteria should be based on the nature of the work to be undertaken, the interests of the firms involved and their vision of how the organization should develop.

The organic quality of organizations comes from deciding which firms to involve, as far as possible on the basis of those firms that like working together. So, in selecting firms it is worth asking 'Which consultants and contractors does the customer like working with?', 'Who are the designers' favourite contractors and specialists?', 'Who are the contractors' favourite designers and specialists?', 'Which designers and contractors do the specialists prefer to work with?'.

An important decision is whether the arrangement should include just one, or more than one firm of each type. The decision will be influenced by the amount of work available and how capital-intensive the necessary technologies need to be. Also, in deciding how many firms of each kind to involve, it should be remembered that it is not ideal to be locked into a single supplier nor a single customer. It is often argued by experienced managers that it is better to have several partnering arrangements running in parallel to retain the incentives provided by competition.

The claimed advantages of partnering with more than one supplier are that there are more good ideas, there is greater security and there is an opportunity to use competition to ensure that all the suppliers keep searching for improvements. However, unless the buyer can sustain a regular stream of work for all the suppliers, the advantages of competition have to be balanced against a reduced commitment to the joint work.

In practice, when a firm has a high proportion of their work committed to only one other firm, it is usually taken as a warning sign. This is commonly interpreted to mean that no firm should have more than 25 per cent of its turnover committed to any one firm. Yet this rule of thumb is commonly broken in strategic partnering. Many firms have too few projects to justify having more than one or two suppliers. Equally, many firms work solely as a subcontractor to a larger firm, getting all or virtually all their work from the one source. Much of the literature on supply chain management accepts that exclusive relationships are necessary and reports many cases in industry generally which continue profitably for the firms involved for many years. However, relationships where one party is heavily committed place large responsibilities on the other firms involved to maintain a real commitment to their long-term success. Firms exposed by being highly committed need to be certain that their partner firms fully intend to sustain the relationship.

Membership problems arise when markets and technologies change. In best practice strategic partnering, firms work hard to maintain the joint arrangements by looking for new opportunities with their partners. However, major changes in markets or technologies can make it necessary to change the membership of a partnering arrangement. This can arise when existing members come to the conclusion that little further improvement in performance is possible. It may be because the current technology has reached maturity and there really is little scope for significant improvement. Or it may be that the assurance of a steady flow of work has made them lazy. In any of these situations the possibility exists that a radically new answer will come from an outside firm prepared to invest in what they see as an opportunity to break into the existing cosy arrangement. This constructive competition is essential for the long-term health of the industry. It helps find the best answers.

A nice example of this kind of initiative arose when three main contractors in a partnering arrangement with Asda, having over some four years reduced the time to construct a new superstore from forty-two to twenty-five weeks, decided there was little room for further improvement. At this point the UK division of one of the large Japanese contractors, Kajima, who were not one of Asda's partnering contractors, invested six months development work looking at radically new ideas. They made this investment at their own risk with no assurance of work from Asda. Their ideas were sufficiently interesting for Asda to award them the contract for a new superstore in Swansea, on condition that it was constructed in less than twenty weeks to the same quality and at the same cost as Asda's existing main contractors would have achieved. The resulting superstore provides an exceptionally attractive place for shopping and, by using unusually well thought out construction technologies, was constructed in just sixteen weeks.

The possibility that radical new ideas will emerge from outside the organization should be taken into account in the formal arrangements between member firms. Rolling contracts are one answer that provides flexibility to change the membership with a reasonable degree of continuity for everyone involved. Rolling contracts usually run for three or more years and are reviewed annually. This gives the firms involved at least two years security. The contracts should provide a minimum call up which is at least sufficient for the supplier to cover overhead costs. This approach allows a long-term relationship to be ended but in a manner that gives time for an exclusive or heavily committed firm to look for other markets or other suppliers. Arrangements of this kind provide a secure base for firms to concentrate on achieving the full benefits of strategic partnering.

The use of rolling contracts is well established in partnering arrangements that involve facilities management activities. Traditionally, maintenance contracts are rolled forward from year to year and the market is tested every few years. Partnering provides, through its emphasis on continuous improvement, a basis for these arrangements to continue for many years. The certainty this provides for the customer and the maintenance contractor has many benefits. The customer can rely on the contractor to understand all the details of their organization and facilities and so be able to work in ways that maximize the value of their work and minimize interruptions to the customer's business. The contractor has the benefit of a reasonably certain workload and secure profits. The same principles apply when continuity is provided by a series of new facilities.

Given continuity of profitable work, most firms will respond by investing in their own future through training, R&D and buying new equipment and systems. This makes them more efficient and so gives the opportunity of devoting time and resources to developing new products and services. This is why effective organizations seek to provide continuity for their member firms because it provides powerful incentives for them to invest in their joint success. Given continuity, best practice includes financial arrangements that enable all the firms involved to earn steadily increasing profits. This is possible because, as partnering leads to better ways of working, so money is available both for higher profits and lower prices.

Creating certainty is a two-way process. Construction firms need to know customers' plans but customers need to be sure they are getting ever better value for their money, at least equal to anything available elsewhere in the market. It is fairly common for customers to tell consultants and contractors the type and number of projects planned for future years, the likely timescales and the terms on which the work will be available. This helps the construction firms involved in a partnering arrangement plan their work with greater confidence.

It is less common for customers to be provided with good information that gives them a similar level of certainty. This could be achieved by consultants and contractors benchmarking their performance in terms of its value to the customer's business. Then they should make sure that they have regular opportunities to present the results to the customer's senior management. This is essential to sustaining certainty because, without regular assurance that strategic partnering is providing benefits, it is all too common for customers' main boards to decide to test the market and a rogue bid is allowed to destroy relationships that are beneficial to everyone concerned. Any threat that this may happen undermines the certainty on which strategic partnering depends.

Continuity cannot be guaranteed because nothing can be absolutely certain in business. This is why there are well established 'get out' clauses in rolling contracts to deal with events outside the control of the firms involved. What is needed for partnering to succeed is clarity over the partners' intentions and consistency in putting them into effect. This should be reinforced with an open exchange of information about the plans and performance achievements of all the firms involved.

Membership criteria

The firms selected to form a strategic partnering arrangement must be competent. This means they understand the leading edge of their own technologies, have well established training programmes and systems to encourage new ideas, and generally encourage innovation and enterprise throughout the firm. It means they deliver on their promises, keep to completion dates, do not make claims and handover new facilities with zero defects.

The firms selected need also to take a long-term view which gives them the confidence to fully commit to partnering. Many firms in the UK still need to move beyond an all too prevalent concentration on short-term goals, too great an emphasis on dividend payments and share prices, and an over-reliance on market-based contracts and confrontational styles of management. Partnering cannot flourish and the industry cannot become efficient when these adversarial attitudes prevail.

In making decisions about the firms to be included in any construction organization, it is important that everyone involved is open and transparent. This means being open about contract conditions with suppliers and subcontractors so firms upstream can be confident that all the benefits of strategic partnering will be available downstream. The point is that many of the ideas for improving performance originate in the experience of people doing the direct design or construction work. Unless their firms are fully involved in partnering arrangements, there are unlikely to be time, resources or systems in place to capture these ideas for the benefit of the whole team. The good ideas will be lost. Therefore, the strategic team needs to ensure that the contract conditions used on their projects are consistent with the strategic objectives and, in particular, that they encourage everyone throughout their supply chains to be committed to continuous improvement.

This kind of well founded certainty is fundamental to long-term efficiency because it allows firms to concentrate on improving their performance. Freed from the need to search for work, devise competitive tendering strategies, haggle over terms and conditions of contract, they can use their best people to lead the search for continuous improvement. Indeed the reduction in overhead costs resulting from the elimination of these competitive, adversarial activities provides important savings.

Firms should be selected on the basis of a broad evaluation of their competence and culture, not just on price. The most highly developed arrangements undertake these broad evaluations at regular intervals. The results are turned into a numerical score that is combined with the firm's price in selecting best value. This means that a firm delivering zero defects, working cooperatively with the customer in finding innovations that help the customer's business, investing heavily in training and having every prospect of still being in business in ten years' time can offer a higher price than a firm doing none of these things and still be awarded a significant share of the work.

Firms that consistently fall below a defined minimum performance should expect to lose the work if they do not improve quickly. Also, there should be opportunities for new firms able to offer extra benefits to become part of the organization. For all these reasons membership should be kept under review to ensure that the firms involved continue to be competent, to play a full role in the work, to be well motivated to do it efficiently and to search for better ways of working.

Project membership

On individual projects, the membership pillar requires all the necessary skills to be available early so the people who make up the project team have every opportunity to cooperate in making their best possible contribution to the project's success.

Effective construction depends absolutely on cooperative teamwork by competent, mature professionals at all levels of responsibility. This means, especially when problems arise, that the people involved are consistently committed to cooperative behaviour throughout the life of the project. Acting cooperatively means taking risks in being open and trusting. It needs courage to concentrate on solving problems as a mutually responsible team. When these new behaviours are not fully reciprocated, it is all too easy for people to slide back into old habits, become defensive and secretive, and begin looking for someone to blame for problems.

It remains the case that the construction industry's traditional methods often encourage adversarial attitudes and confrontational behaviour. It is not easy for people who have learnt how to survive in the traditional industry suddenly to change; they often need training in cooperative behaviour. This can begin at workshops but needs to be reinforced by short courses and encouraged by consistent support from senior management. It has to be recognized that not everyone can develop adequate skills in cooperative behaviour. When it is clear that an individual is unable or unwilling to change, the project team must either replace that person or, in exceptional cases for someone unusually talented, decide to make special provisions. In either case, the project team must act when individuals are found to be behaving in a way that does not take other peoples' interests into account.

Investment in training and coaching are essential to develop the skills required for partnering. Bennett and Jayes (1998) describe case studies which show that many firms realize that staff at all levels should be better at the core skills of process analysis, work measurement, planning techniques, the use of control charts and basic statistical techniques. These firms are beginning to organize training courses for all their staff to encourage a consistent approach throughout the organization.

This is a new development in the construction industry's flexible labour markets where individuals have traditionally had the primary responsibility for their own development. Never-the-less firms who want to build long-term futures for themselves and their staff are beginning to encourage and reward people who make the effort to acquire new skills. Partnering encourages this because it provides a framework in which individuals and firms all benefit from enhanced skills. The membership pillar requires that attention is given to this continuing need for learning so that everyone gets the full benefits from partnering.

Equity

The equity pillar seeks to ensure that everyone in the organization is treated fairly in ways that mean they do not need to waste time worrying about money issues. The key features are listed in *Figure 4.9*.

An important product of this pillar is a strategic business case for the construction organization. Like all other commercial organizations, construction organizations are funded by profits from providing customers with products and services. However, a difficult issue which goes beyond this is long-term development work, especially its funding and the ownership of any benefits it produces. This is potentially a problem because most construction work is funded project by project but efficiency requires firms to invest long term.

EQUITY
Strategic business case
Long-term development
Agreed allocation of costs
Fair rewards
Long-term funding
Continuity of learning
Fair budgets and prices
Joint liabilities

Figure 4.9: Key features of equity

Customers with regular programmes of building work often fund long-term development work. Also the bigger firms in the industry, mainly contractors plus a few consultants, undertake long-term development work.

However, this is mainly very specific, in support of a customer with a regular programme of work.

Bennett and Jayes (1998) report several examples where strategic partnering teams have formally agreed to fund development work independently of any specific project. The usual arrangements are for each member of the strategic team to fund a proportion of the costs of long-term development work undertaken jointly. One typical example allocates 25 per cent of the costs each to the customer and main contractor, and 12.5 per cent each to the architect, structural engineer, services engineer and quantity surveyor.

There are advantages to the involvement of specialists and manufacturers in strategic development because these sectors include many of the largest firms in the industry with the financial strength to act long term. It is important also because the industry needs to build up efficient supply chains. Where specialists and manufacturers are included they too are allocated an agreed share of the costs.

Firms can look beyond their own resources for the financing of long-term development work. Government has been instrumental in providing pump priming for successful joint initiatives that subsequently rely on industry funding. In the UK, these normally include links between industry and universities or other research bodies. However, it would be unrealistic for the construction industry to rely on Government funding to provide much beyond initial costs of long-term development.

Ultimately, firms must rely on shareholders or customers to provide the resources they need. The industry's low profit margins and poor reputation with financial institutions make it unlikely that construction firms could use equity capital to raise substantial amounts of long-term money. Banks provide another possibility but it would take a major change in culture for them to raise their long-term stake in construction companies. One possible source of shareholder funding, at least for the bigger building companies, is international construction companies seeking a presence in local markets. A number have already bought into UK companies and have brought with them a greater willingness than local firms to think and invest long term.

The other primary source of finance is the industry's customers. There are two promising ways in which they may provide more long-term finance. The first is savings produced by strategic partnering. The main issue for construction firms is the extent to which they will be able to hold onto a fair share of the higher profits. Many leading customers are determined to reduce their costs and construction firms are willing to settle for greater certainty of profits and let their customers take the lion's share of the additional rewards. This makes short-term commercial sense but provides no basis for the industry to invest long term. Construction firms will have to overcome these traditional short-term attitudes if they are to get the full rewards that strategic partnering can provide.

The second promising source of funding for long-term development is provided by private finance initiatives (PFI). A number of important innovations have been developed, including long-term bank loans already extending over twenty years and various forms of bonds. Some projects involve different forms of finance for distinct stages, typically bank loans to cover the risks in the construction stages and then, once a facility is up and running, switching to a long-term bond.

PFI schemes often involve construction organizations in designing, constructing and running a new hospital, school, prison, university hall of residence or other public facility for up to thirty years, in return for an annual payment from the organization that needs the new facility. These schemes invariably include several firms contributing distinctive skills and resources. PFI requires firms to think long term. Life-cycle costs have to be taken into account. Quality and maintainability issues all have to be considered in new ways. Even more fundamentally, designers and financiers have to understand how buildings are used and are likely to be used in the future. So, PFI is forcing changes in the way all sectors of the construction industry think and act. These changes are consistent with the demands of the new paradigm. A key part of making this possible is the income stream generated by PFI which provides reasonably secure long-term finance.

So there are various possible sources of financial support for long-term development work. Given the necessary finance, the key practical issue in ensuring that development work is effective is creating continuity. Without continuity, learning points get swamped as people are forced to deal with the same old problems and so become discouraged and disillusioned. Providing continuity requires action from the top. In considering if it is worthwhile making the effort to provide continuity, strategic teams need to remember that it provides many benefits. Many of these result from giving everyone involved that sense of financial security which motivates them to focus on doing their best work in pursuit of the agreed strategy.

Project equity

The best partnering arrangements enable everyone involved in individual projects to concentrate on developing and achieving agreed objectives without having to worry about whether they will get paid for what they do. The way rewards are handled should give all the firms involved substantial confidence that they will get a fair return. This is energizing and provides the best basis for high efficiency.

Therefore an important preliminary aim should be to provide a fair return in normal business terms for everyone involved. In agreeing how rewards are to be dealt with, the general aim should be to align everyone's interests with fully meeting the customer's objectives. This means arranging the payments to each construction firm so their profits increase directly with the value the customer receives from the project. This should provide greater returns than traditional approaches because partnering exists to find better answers, including finding savings that can be used to fund real incentives for the whole team.

The worst situations in any project arise when some of the firms involved are losing money. Therefore, the objectives agreed at the outset must not create the risk that some of the firms, even if they contribute their very best efforts, will lose out relative to the others. Partnering depends on there being a genuine alignment of interest in which rewards and risks are shared in ways that take account of the commercial realities of the firms involved. There is no point in pressurizing small firms into accepting a big risk; it will dominate their thinking and prevent them doing their best work.

Currently best practice is for the customer to pay the firms they are partnering with an agreed sum which provides a fair profit and contribution to their fixed overheads plus all their properly incurred direct costs. This empowers people to concentrate on doing whatever is in the best interests of the project and to consciously eliminate unproductive activities. When they know they will be paid for whatever they do, as long as they are honestly seeking to add value for the customer, people are prepared to try new ideas, join in task forces, set up and experiment with prototypes and generally do whatever is needed to find the best answers. Working in this way requires open book accounting, purposeful cost control and tough auditing. This approach provides a basis for firms to totally alter the way they work. Instead of defending their own narrow interests in an adversarial manner, a fair approach to the financial arrangements causes them to think and act as a team. They accept that there is a fixed amount of money available and they are all jointly responsible for providing the best possible value within that sum.

A key part of this approach is some means of assuring the customer that the agreed price provides value for money. Benchmarking based on projects for the same customer is an effective approach but where this is not possible, using similar projects for other customers provides an effective alternative to competitive tendering. The overall aim is for customers to set budgets and for construction organizations to establish prices that fairly reflect all their interests. The process of establishing budgets and prices may mean the parties have to negotiate at length on the basis of an open exchange of information until they reach an agreement they can all regard as fair.

To keep things administratively tidy, the agreed arrangements for dealing with money must reflect the form of procurement used, so there are no contradictions. The choice of contract is often seen as an issue for people who want to retain the powers that traditional forms give them if partnering fails. They see this as a safety net to be used when problems arise. When regarded in this way, traditional forms of contract can undermine partnering because they make it too easy to fall back into adversarial approaches justified by procedures in the contract. It is therefore important for any contract used to reflect non-adversarial ways of working. The NEC Engineering and Construction Contracts are good examples of what can be done to foster a cooperative approach to construction.

In addition to the contractual provisions, all the formal legal positions should be taken into account in agreeing the reward structure. One arrangement that helps overcome some of the complex patterns of liabilities that can frustrate partnering is to set up project-based professional indemnity insurance. This helps everyone concentrate on the best interests of the project as a whole rather than worrying about their own legal positions.

Integration

In order for the firms making up a single construction organization to be really effective they need to integrate their procedures, systems and cultures. Kanter (1989) describes five levels of integration that need to exist between firms for them to achieve the most productive relationships. The first three deal with the need for integration at different levels of work. Kanter's identification of three levels is grounded on the US management paradigm, never-the-less it identifies the issues that need to be dealt with in order to achieve integration. The key features are listed in *Figure 4.10*.

INTEGRATION
Strategic objectives
Tactical procedures
Operational systems
Interpersonal relationships
Cultural attitudes
Cooperative behaviour
Trust and honesty

Figure 4.10: Key features of integration

Strategic integration deals with separate firms' goals and objectives and identifies changes that are necessary. The more information communicated and the more contact between different firms at this level, the greater the possibility that the organization will evolve in cooperative rather than conflicting directions. In construction, strategic integration often begins with informal meetings between senior managers of firms who have worked together in the past and who want to consider a more permanent relationship. These meetings deal with big issues and play a large part in ensuring progress in integrating the other levels. These are, of course, central issues for strategic teams.

Tactical integration brings together middle managers and professionals operating at equivalent levels in the separate firms. They develop plans for specific projects or joint activities, identify organizational or system changes and flows of information that will better link the companies.

Operational integration provides ways for people carrying out day-to-day work to have timely access to the information, resources or people they need to accomplish their tasks. The aim should be to build up an attitude of doing everything necessary to meet targets. This includes solving problems as they arise, recognizing new opportunities and providing feedback on their performance.

Interpersonal integration provides a necessary foundation for effective work. Many people need to know those in other organizations personally before they will make the effort to exchange technology, provide access to customers or participate in joint teams. Good interpersonal relationships provide this kind of confidence and help resolve conflicts before they escalate into disputes.

Cultural integration requires everyone involved in the organization to have sufficient communication skills and cultural awareness to explain their views in other people's terms. This helps bridge differences. Cultural integration is encouraged by managers demonstrating interest and respect for other people's points of view. This builds a depth of understanding and goodwill that helps smooth over organizational differences.

Life-cycles of integration

A key responsibility of strategic teams is to keep analysing joint processes to maintain continuous improvements in their performance. In doing this it should be recognized that any given type of work moves through a life-cycle. This is commonly illustrated in the form of a sigmoid curve in which initially work is poorly understood and undertaken inefficiently. Then as understanding grows so efficiency increases at a gradually accelerating rate. However, there are limits to these increases for any type of work and so a point is reached where improvements in efficiency slow and work reaches a mature state. Further improvements in performance are difficult to achieve and are relatively small. The maturity stage is a signal for managers to look for a step change to a new type of work with the potential to deliver future improvements in performance. Step changes are necessary to ensure the long-term survival of organizations but putting them into effect is often at the expense of an initial reduction in efficiency. So, changing from the mature stage of an old technology to the initial stage of its new replacement inevitably causes uncertainty for many of the people involved. The new paradigm's reliance on open communication is especially important at such times if people are not to feel threatened by rumours of change.

The level of integration that can be achieved is influenced by each of the three stages. This is well described by Handy (1994) and illustrated in *Figure 4.11*.

The life-cycle of integration begins with an initial learning stage because the new type of work is not well understood. There is little relevant literature or research. The particular skills and processes needed to undertake it are not well developed. The performance levels delivered by the product are not very predictable. The support services that customers will demand are not fully defined. Therefore, the nature and sequences of work required to produce a new product have to be determined on a one-off basis. Work in the initial learning stage can be characterized as non-standard, experimental trial-and-error requiring high levels of creativity. It is essentially problem-solving work using what Schon (1983) calls reflection-in-action. This means trying a promising idea, reflecting on the results it produces, thinking of an improvement, trying it, reflecting on the results, and so on.

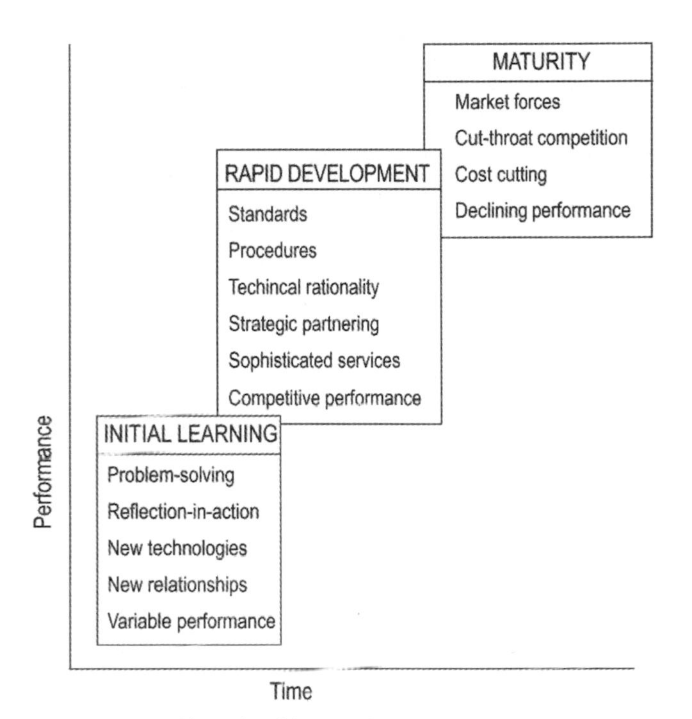

Figure 4.11: Life-cycle of integration

This way of working requires considerable cooperation between the separate firms involved in the work. In many cases they are working together for the first time. Thus the initial learning stage is characterized by new relationships and changes to existing relationships. It will often be the case that the firms need to change attitudes and their own internal procedures in order to support cooperative ways of working with other firms. As a result performance is initially inefficient and variable. However, this groundwork is essential to provide the basis for the real integration that comes in the next main stage of the life-cycle.

This is the rapid development stage in which work becomes highly rationalized and uses increasingly well integrated working methods. There are significant advances in the understanding of the work. Research is undertaken and published. The best methods are brought into the education and training of established professions and crafts. The performance that will be delivered by the products, and the nature and sequences of work required to produce them, are becoming established. Standards appear and are developed and improved. Well developed organization structures and processes appear and become increasingly formalized.

These developments take time and committed resources and so long-term relationships play a large part in the strategies of successful organizations. It is in the rapid development stage that strategic partnering delivers its full benefits. As a result, performance becomes significantly more stable and high levels of efficiency are achieved.

The successes of the rapid development stage lead to the working methods becoming well established. They provide the basis for professional and craft skills and the kind of knowledge that Schon (1983) calls technical rationality. The performance that will be delivered by products, and the nature and sequences of work required to manufacture and construct them, have been researched, are well described in widely available literature and so work consists of applying well developed methods. Work becomes highly differentiated, and efficient. Results are predictable within narrow margins. Organizations responsible for similar types of work compete for the available market. The support services provided to customers become increasingly diverse and sophisticated as firms compete to differentiate their, by now, very similar products.

So the life-cycle has entered its maturity stage in which to a large extent all firms use the same highly efficient methods to provide competitive levels of performance, otherwise they go out of business. The long-term relationships that grew up during the rapid development stage provide fewer benefits and tend to be replaced by a reliance on market competition as buyers see opportunities to find lower prices. Competition becomes more cut-throat, firms go out of business and a gradual decline in performance sets in as the most talented and enterprising individuals and firms make a step change into a new kind of work. This signals a return to a new initial learning stage and the start of a new sigmoid curve.

Much recent management literature suggests that the rate of change in markets and technologies has become so rapid that organizations have to make step change after step change in order to satisfy ever changing customer demands. They rarely have time to move into the rapid development stage of the sigmoid curve and so have to remain flexible, uncommitted, ever-changing virtual organizations. There is undoubtedly some truth in this, but it remains a responsibility of all the teams involved in construction to create areas of certainty that allow integrated processes to develop. It is in these areas that partnering will be effective and long-term improvements in efficiency will be achieved. The more that key external organizations, especially customers and suppliers, are integrated with the construction organization, the easier it will be for construction teams to create sustainable areas of certainty.

Cooperation and trust

Integrated processes make it much more likely that cooperation will become the vital guiding principle in interactions between all the people who make up construction organizations. As people develop cooperative habits, they learn to trust each other. Trust is difficult for many people because of their experience of traditional business methods. In the past, they have been forced into narrow adversarial and competitive situations that provide no outlet for their creative talents. This is one major reason why many firms are inefficient.

Trust means acting on the belief that the other party will do as they have said. Trust depends on honesty which, in practical terms, begins by not promising more than you can deliver. Covey (1989), in an important review of successful behaviour, describes trust as the highest form of human motivation. It brings out the very best in people. But it takes time and patience to create, and so strategic teams have to keep working at building trust throughout their organizations. A vital part of this is for all managers to treat their subordinates exactly as they want their best customers and suppliers to be treated.

Trust has to exist inside firms which means that authority is delegated and people are empowered to work cooperatively. Trust provides a secure basis for open communication, mutual learning and real creativity. Trust allows teams to focus on interests rather than on personalities or positions. Trust promotes openness and encourages people to put their cards on the table. A sign of trust between two parties is that they are both committed to try to understand each other's interests and to work together to find answers that provide more than either of them expected.

Covey (1989) concludes that trust-building is the single most effective investment in business relationships. It gives people, teams and organizations the freedom to do their best work. Practical benefits that he found included fewer inspections and audits, fewer decision levels and faster communications. The most important benefit of trust is that it produces the genuinely win:win situations that are the principle aim of the integration pillar.

Project processes

The new paradigm forces fundamental changes in the way individual projects are tackled. They are no longer essentially discrete sets of actions with a clear beginning and ending, rather they should be seen as components of an ongoing long-term activity driven by feedback. The key features are listed in *Figure 4.12*.

PROJECT PROCESSES
Feedback driven
Standards
Procedures
Development by task forces
Mainstream efficiency
New-stream creativity
Analysis of customer's needs

Figure 4.12: Key features of project processes

This means project teams have less discretion in how they tackle individual projects than has traditionally been the case. Instead of searching for individual answers, they contribute to a long-term strategy put into effect through an ongoing series of projects. This alters the traditional work of all the construction professions.

Traditionally, project work demanded decisive action in solving problems based on professional experience and judgement. Strategic partnering requires project teams to concentrate on applying the current best answers embodied in procedures and standards. Thus, project teams carry out predetermined sequences of actions determined by careful analysis and research undertaken outside of individual projects. This is likely to be the most fundamental change for people used to the traditional methods of the UK construction industry. The detailed implications for all construction teams are described in Chapter 6.

A central feature of the new approach is that development and innovation are undertaken by task forces set up by the strategic team to improve the current best answers. New ideas are tested and improved through their carefully controlled use on individual projects. Problems thrown up by these tests are tackled jointly by the project team and the task force working together and the results fed back to the strategic team. Once established and embodied in procedures and standards, new ideas are used on all the organization's projects.

Procedures and standards establish design details, predetermined roles, patterns of meetings, flows of information and planning and control systems. In this way they support an almost automatic way of working which enables project teams to concentrate on dealing with the few unique features of their individual project. In many cases these mainly concern groundworks, external works and planning requirements that influence the external appearance of buildings. Even for those elements, the most highly developed practice provides standardized approaches to dealing with most project situations.

This approach depends on relevant procedures and standards being in place so that project teams can be assembled and can carry out individual projects virtually automatically. Each firm needs to have well trained teams able to apply the agreed procedures and standards as part of an integrated process. Professional roles inevitably change as best practice identifies which skills and knowledge add value. For example, quantity surveyors often emerge as particularly effective at analysing, measuring and improving processes, once they break away from traditional routines geared to adversarial contract conditions.

To allow the new paradigm to be applied, procedures and standards must provide for mainstream and new-stream projects. To achieve this, strategic teams decide the distinct products and services their organization will offer to customers. Then they ensure that each product or service is supported by well developed procedures and standards. As new projects become available, they decide which of their predetermined approaches should be put into effect. Finally, a project team of people experienced in the selected type of product and services is set up.

It is often unclear which answers will best suit a new customer. In these circumstances, the strategic team should establish an embryonic project team of very experienced professionals to work with the customer in developing a mission statement and brief. Once it is clear what the project's objectives should be, either the embryonic team is turned into a full project team, or a new team is set up comprising people experienced in the required work. The outcome will often be a mainstream project but occasionally a new customer is best served by a new-stream project. It is vital that when this happens the customer has clear information about the costs and risks involved and these are fully discussed by the project team so that everyone understands the nature of the work.

The overall aim of the project processes pillar is that basic work is undertaken efficiently in the long-term interests of customers, construction firms and the communities affected by their work. These interests include ensuring that construction helps build sustainable communities without damaging the world's environment.

Benchmarks

A key aim of construction organizations guided by the new paradigm is to deliver continuous improvements in performance over a series of projects. The benchmarks pillar deals with the need to measure performance to ensure that these continuous improvements really are being achieved. The key features are listed in *Figure 4.13*.

Benchmarks help prevent long-term relationships becoming cosy and sloppy by providing objective measures of the performance achieved year-on-year. In this way they play an important part in convincing the firms that make up a construction organization that strategic partnering involves fewer risks and delivers bigger rewards than traditional competitive methods.

BENCHMARKS
Continous improvement
Measured performance
Challenging targets
Reliable information
Value-for-money
Project comparisons
Norms and key targets

Figure 4.13: Key features of benchmarking

Measurement is essential to cooperative efficiency because it provides data which help strategic and project teams set targets that deliver continuous improvement. Given good data, it is hard for people to allow their work to become cosy, inefficient or sloppy. The main argument in favour of competition is that it establishes the market price for any given product or service; rigorous measurement can provide the same information without creating the adversarial framework that competition engenders.

The benchmarks used should be capable of allowing comparison of costs, times, quality and service with those being achieved by other firms in the same market sector. In this way customers can be confident that they are getting good value for their money. This is what competitive tendering provides but at considerable cost in terms of unproductive activities. Benchmarks provide a robust alternative to competitive tendering by giving customers objective information about the options available.

Benchmarks used to measure products and services should be simple, robust, widely understood and deal with whatever is important to customers. It remains the case that they need to demonstrate cost competitiveness because this is a factor in most customers' decisions. Good examples of effective benchmarks include: cost per square metre for buildings, square metres of building constructed per day, total project time from inception to occupancy, number of defects at handover, complaints from users about their building, customers' views on the new facility, and life-cycle costs.

As well as measuring performance in terms of the products and services delivered to customers, benchmarks should help drive continuous improvement by measuring the performance of the overall organization, project teams and firms. In deciding on the benchmarks to measure the performance of construction organizations and their various parts, it is important that they should be consistent with a common sense, practical view of success. Benchmarks would be rapidly discredited if they suggested that the organization was doing well but several of the firms were losing money and users were not happy with their new facilities.

It is often helpful to measure workers' perceptions of the things that influence their performance. These could include their perceptions of induction courses, the effectiveness of meetings, safety training, safety provisions, site facilities, clarity of information, level of interference from other teams and similar matters.

Other interesting approaches that may help in the search for robust measures include: measuring how long it takes to halve the time of individual processes; measuring how long it takes to halve the number of defects; asking the customer how effective the partnering team is at listening to their concerns and needs; and how effective they are at responding to what they hear. It is often helpful to discuss what 'winning' means in the context of the partnering arrangement and then developing benchmarks that measure the resulting criteria.

It is commonly argued that since all construction projects are unique, no useful measures of project performance can be devised. However, best practice demonstrates that it is possible to have useful measures that draw on the experience of other projects. Bennett and Jayes (1995) suggest that measures should be turned into a numerical ratio or percentage so that the units of measure are not project-specific. Then projects can be compared directly and realistic targets set.

Key targets are often set at the strategic level as part of the aim for continuous improvement over a series of projects. For the key targets to be effective, they need to be based on well established benchmarks. In setting the key targets, it should be recognized that there is an established normal level of performance that all project teams should be able to achieve. Key targets relate to improvements beyond the norm that the team are being asked to achieve.

Project processes should include measuring performance against the norm and against the key targets. This should not need extra resources nor be expensive. The measurements should concentrate workers' attention on their own performance to help foster the attitude that every worker is responsible for the project's performance.

Benchmarks demonstrate that continuous improvement is being achieved and so help convince customers and senior management to continue to support the organization's use of strategic partnering. Without robust measures, especially of value for money, few customers will be persuaded to use partnering. This is especially an issue with inexperienced and one-off customers. They need convincing evidence of the benefits and the low level of risks involved. Objective measures of performance, benchmarked to industry norms, provide a powerful part of this evidence.

This is why it is essential that the industry develops robust measures of improvements in performance delivered by partnering. It is helpful if these benchmarks are expressed in terms that can be related to the balance sheet or, even better, to the share price. This is so that customers' senior management have clear evidence of the advantages of partnering in the language they use daily in making business decisions.

Although it is important to express performance measurements in financial terms, within industry generally there is growing concern that traditional financial measures do not tell the whole story about performance. Many firms are now understanding the need to identify realistic measures of all aspects of business performance, including their impact on the environment and the community.

RSA (1995) reports that most major companies in the UK recognize the importance of long-term relationships as a source of competitive advantage. There is a growing understanding that teamwork and long-term trust are more effective than power-based, adversarial relationships. However, in practice, there is little serious measurement of relationships and, unless the effects are seen in bottom-line figures, there is an inherent risk that efficient relationships may be dissolved. The RSA report concludes that, until other measures are treated as seriously as financial ones, the crucial importance of building effective relationships will continue to take a back seat when decisions are taken at board level. In other words, benchmarks are crucial in putting the new paradigm into practice.

The UK's Construction Best Practice Programme, which is supported by central Government, has recognized the importance of establishing widely used benchmarks. A big step towards this has been the publication of Key Performance Indicators that measure project and company performance.

Feedback

Feedback is essential for control and for continuous improvements in performance. Feedback is knowledge available to a system about the effects of its actions on its environment. The construction industry has many good ideas that are lost because there is little systematic feedback. Lessons should be captured so that they can be applied on future projects. This is commonly called positive feedback, because the organization responds to the feedback by continuing with the same actions. Negative feedback is also vital in identifying problems and defects and ensuring that they do not re-occur. Negative feedback is that which causes organizations to respond by changing its actions because they are causing performance to deviate from acceptable norms. In these two distinct ways feedback drives control loops that enable construction organizations, project teams, task forces and firms to steadily improve their performance and identify problems quickly. The key features are listed in *Figure 4.14*.

FEEDBACK
Control loops
Continuous improvement
Communication channels
Designed for individual needs
Workshops
Mainstream automatic control
New-stream flexible control

Figure 4.14: Key features of feedback

For feedback to play its full role, firms involved in strategic partnering need to develop procedures and standards that systematically capture best practice as it emerges from project teams or task forces. It is especially important to focus on understanding and improving joint processes because strategic partnering often provides its biggest benefits by improving the interfaces between teams. Feedback from other organizations about their experience of working with an organization can be especially revealing and often provides the basis for significant improvements to working methods, products and services.

Feedback is used in this way to improve efficiency in many different situations. Indeed, the efficiency of all controlled systems depends on accurate and relevant feedback. This general experience tells us that construction should be seen in terms of controlled systems; not doing so is one of the key weaknesses of traditional, project-based approaches. The widespread perception that all projects are unique has limited the use of feedback. Consequently mistakes are repeated, good ideas are lost and the industry's overall performance falls behind other modern industries. The answer is for construction to set up effective feedback systems as a matter of urgency. Agreed benchmarks determine what feedback should be collected and how it should be analysed. Then there need to be clear communication channels that take the resulting information to wherever it is needed to influence decision making.

Information technology has revolutionized the ease with which feedback can be collected, analysed and distributed. For example, some construction organizations routinely install video cameras on construction sites so that anyone in the organization who needs to know what is happening can simply take a look. This provides feedback in real time so that when a problem arises, people in remote offices or on another site can see the situation and provide advice or contribute to decisions. In this way information technology both speeds up and improves the quality of decision making.

An issue that warrants careful thought by the strategic team in making best use of feedback is who should receive feedback, in what form and what they are authorized to do with the results. In designing systems that replace human communication with information technology, it is vital to consult the people involved about the form in which the feedback results are required. Some teams are happy to have no more than a list of exceptions to planned progress; others want to see the trends; yet others are more comfortable if they can see the detailed data which produced particular results.

An important part of feedback comes from workshops on individual projects, especially the final workshop. These should review the performance of the project team and produce succinct reports for senior management on the key achievements, problems and new ideas. In reporting to senior management the benefits should be expressed as hard facts in bottom-line terms. They should be illustrated by examples which show how the team has cut out waste. They should also emphasize and describe the benefits for individuals, teams and firms if they continue to be involved in the overall construction organization.

Effective feedback systems are an important element in adopting the new paradigm. They help construction organizations establish virtually automatic control for mainstream projects and flexible control for new-stream projects. Such sophisticated control systems, using the latest information technology, are becoming a matter of routine in the management of leading firms in most industries. Construction has a great deal of ground to make up to match this best practice. A crucial step is to establish effective feedback systems because, as *Figure 4.4* suggests, feedback turns the other six pillars into a controlled system capable of guiding construction in its adoption of the new paradigm.

The seven pillars in practice

The seven pillars of partnering provide a framework designed to make construction simpler and more certain than when traditional approaches are used. They enable competent people throughout the industry to concentrate on producing quality and value. They encourage the long-term development of construction organizations. They help construction aim at sustainable development, shape major changes to its markets and technologies, and be responsive to change.

These important benefits require traditional construction teams at all levels to rethink their organization structures and processes from basic principles. The next two chapters describe this rethinking in careful detail in the hope that it will help all those involved in the industry understand the practical effects of the new paradigm.

Structures 5

Traditional structures challenged

Hierarchy is a fundamental part of traditional management thinking. As Belbin (1993) puts it, managers have traditionally believed that work gets done only when someone has the power to order a subordinate to do it. However, this view is now widely challenged as traditional hierarchical structures increasingly fail to deliver improvements in efficiency. Belbin's research into effective teamworking shows that rules and regulations imposed from above restrict and limit initiative. As people are better educated and informed, they want to make their own decisions. So hierarchies are subverted as workers introduce their own ways of doing things, despite the rules.

Many organizations have responded to these challenges by reducing the number of levels in their management hierarchies and encouraging staff to develop relationships that cut across formal boundaries. The management research by Peters (1987) is one well know example that encouraged organizations to see their structures as networks of multi-skilled, self-managed work teams tied together by business ideas, customer relationships and information technology. Peters urged teams to talk to customers to find out what is really needed and to work with them to solve problems and find better answers. Similarly he advised teams to work with suppliers because they have specialized knowledge and skills and are a rich source of new ideas and good questions. As more decisions are made in basic work teams, fewer levels of management are needed and hierarchies become flatter.

These changes, which represent a new paradigm of human organization, have been reinforced by the development of information technology which deals with much of the routine collection and processing of information previously undertaken by layers of middle managers. The wider understanding of the networks of relationships that shape modern organization structures has coincided with the development of information technology networks. There is little doubt that these changes are linked and that they reinforce each other.

It is now increasingly evident that despite widespread down-sizing, many major companies continue to grow. What has happened is that senior management, having first made its various business units efficient by restructuring, faces pressures from shareholders and their advisors to continually increase market capitalization by expanding the business. In many cases this means that previously independent business units are linked by adding higher levels of management that are needed to run global businesses. In other words, the application of the new paradigm initially leads to fewer levels in management hierarchies. This helps improve efficiency, which increases profit levels and enables businesses to grow. This, in turn, requires additional higher levels of management to deal with the wide ranging and very long-term decisions faced by global businesses.

Construction firms provide few examples of these trends that are helping other industries be more efficient and innovative. Managers responsible for construction need first to understand how structures are influenced by the new paradigm and then put the theory into practice. Therefore this chapter describes the principles of structure derived from the new paradigm and then discusses the practical implications for construction.

The most important point to bear in mind is that rethinking construction in this way, as the Egan Report (Construction Task Force, 1998) explains, makes work simpler and more efficient. These benefits are substantial and they are vital for the construction industry's future health. They result from adopting the new paradigm. This link provides more than sufficient justification for the following careful description of the principles of structure derived from the new way of thinking.

Principles of structure

The fundamental principle of structure provided by the new paradigm is that construction work is most effectively undertaken by networks formed by people and the relationships between them. Individual genius is, of course, important in thinking new thoughts, making connections that no one else has recognized, and seeing visionary futures. In construction at least, and probably in most fields of human endeavour, genius requires support teams to turn new ideas into practical reality. More generally for most ordinary mortals, outstanding performance depends on being part of a well established team. The following description deals with this normal situation in which new kinds of facilities, better designs, new technologies and more efficient methods are produced by teams connected together in networks. The networks of teams form project organizations, construction organizations, firms, communities and other organizations that provide the structures which enable talented individuals to do their best work. Some individual genius in the future will rewrite this principle but, for the moment, construction produces its best work in organization structures formed by networks of people and the relationships that grow up between them.

People

The people that make up these networks have skills and knowledge relevant to construction. This comes in many forms and a rich language has grown up to describe the perceived differences. Terms in common use often conjure up mental images of specific people who have impressed us. However, more generally, they refer to two factors: technological competence and levels of work.

Technological competence

The first factor implied by many of the terms used to describe people in construction is competence in specific technologies. Common terms include project managers, architects, engineers, piling specialists, bricklayers, steel erectors, cladding fixers, carpenters, electricians, painters, estimators, construction planners and construction managers. These few examples illustrate the broad range of technologies required by modern construction which include the techniques involved in financing, designing, planning and controlling work, as well as its direct execution.

An important trend in technological competence is multi-skilling. This means mastering a basic technology and knowing how to plan, control and improve its established methods and techniques. It often also includes learning related technologies so teams can undertake broad tasks. This is very different from the approaches used in traditional production technologies where work is systematically simplified and workers are expected to undertake fully defined and very narrow tasks repetitively. These approaches were carried to extremes in mass production factories by Henry Ford and subsequently by many other firms.

As the human costs of mass production became evident, multi-skilling was introduced in many industries and this has allowed first-line workers to be given considerable responsibility for managing their own work. This includes a responsibility for increasing their own efficiency and looking for improvements in the products and services they produce. Modern technological competence means workers have the skills and knowledge needed to make decisions about the management of their own work in pursuit of objectives and targets they helped establish. Where construction firms have introduced multi-skilling, they are finding that it helps improve productivity and quality.

Levels of work

The second factor implied by many of the terms used to describe construction workers is competence at a particular level of work. Commonly used examples include labourers, semi-skilled workers, craftsmen, foremen, technicians, specialists, professionals and managing directors.

Jaques (1989) describes in great detail the implications of people having different levels of competence. Essentially he sees the differences between people in terms of their ability to handle abstract concepts and deal with long gaps in time between decisions and outcomes. This means that low-level work deals with direct physical work in which outcomes are immediately apparent. High-level work deals with abstract, generalized ideas in which outcomes become apparent only after many years.

Levels of work differ in the number of variables that have to be considered, how clearly they can be defined and their rate of change. Jaques describes this as task complexity. New problems often involve high-level work because the variables involved are ill-defined and changing. Existing knowledge and skills are inadequate to identify the best answers and the required technologies have not been developed. As a problem is studied and the variables understood, good answers are found and reflected in organizations' structures and procedures. Then workers no longer experience that same level of problem complexity because good answers are readily available. The task has been simplified and therefore requires a lower level of work. It is only when task complexity is reduced that tasks can be undertaken successfully by people with lower levels of work skills and knowledge. Eventually, of course, the work is done by machines.

Jaques applied his understanding of levels of work to traditional management hierarchies. The new paradigm uses ideas about levels of work in a simpler way. Jaques' approach identified four distinct ways of working and four orders of complexity in human thinking; so his complete scheme consists of four ways of working with four orders of complexity to provide, at least theoretically, sixteen levels of work.

The four distinct ways of working, as illustrated in *Figure 5.1*, describe increasingly complex patterns that we construct in our minds and use to draw conclusions about problems. The first level is what Jaques calls discrete primary sets of information. These are pieces of information that we believe are related. Typically, primary sets of information are made up of things that we find interesting for some reason, their behaviour, and courses of action that help us achieve our objectives when the things behave in particular ways.

The second way of working involves a number of primary sets of information connected with each other in series. Outputs cannot be fully prescribed, information needs to be assembled and interpreted to solve problems. The general pattern of appropriate methods are known but the specific approach results from reflecting on what is occurring, anticipating potential problems and diagnosing actions to prevent or overcome them.

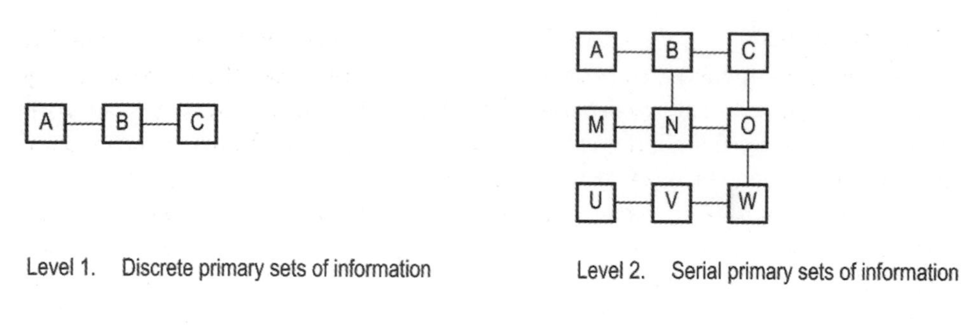

Level 1. Discrete primary sets of information

Level 2. Serial primary sets of information

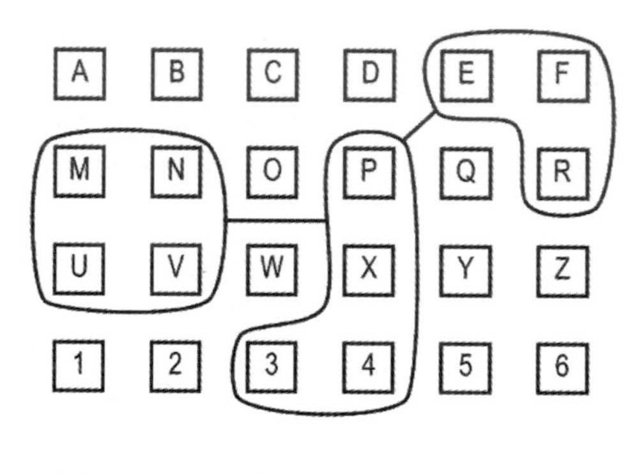

Level 3. Partial secondary sets of information

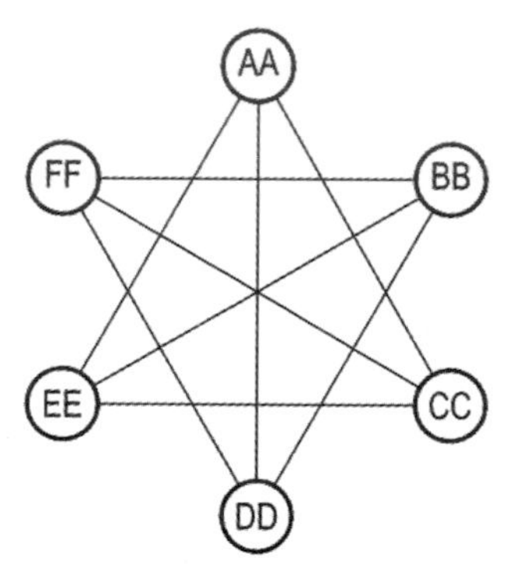

Level 4. Secondary sets of information

Figure 5.1: Four distinct ways of working

The third way of working involves partial secondary sets of information. These arise because there is such a mass of detailed information that it has to be thought of in terms of categories, some of which can be ignored for the time being, to allow attention to be concentrated on the exceptional, critical or most interesting.

The fourth and final way of working involves secondary sets of information. Faced with an overwhelming amount of detailed information, workers generalize it into sub-sets which are used as items of information. It involves balancing several distinct activities, each pursuing their own emerging pathways, that are all interconnected. Resources and priorities are balanced as the separate work makes progress and encounters problems. Targets are long-term and deal with improvements in performance or growth over several years.

People normally develop higher-level skills and knowledge as their experience grows. When they are competent at level four work, further progress requires a new way of thinking which marks the change from one order of complexity in human thinking to a higher one.

Jaques' first order of complexity is thoughts and words that relate to tangible things. This is the thinking of children and refers to physical things that can be pointed out or, if not physically present, have been seen recently and are assumed still to be there. Examples include that door, that drawing, that church and that road. People normally make the transition to the next order before they begin work.

The second order of complexity is symbolic thoughts and words which are used by most adults. Words no longer have to refer to specific tangible things. They are symbols that represent categories of things and can be used as if they were the things. Thus the word door does not necessarily refer to any specific door but to a generalized concept of the characteristics that doors have in common.

The third order of complexity is intangible thoughts and words. These are abstract concepts that refer to other thoughts and words and so refer to tangible things at more than one remove. Examples of intangible concepts include newly emerging markets, culture, values, labour markets, asset values, foreign exchange and management research.

Jaques fourth order of complexity is universal thoughts and words which is the province of what is ordinarily regarded as genius. It deals with types of communities, systems of ethics, the basis of cultures and radically new theories. Examples include capitalism, democracy, equality, freedom, free trade, human rights, competition and cooperation.

The levels of work that need particular attention in forming structures are where the nature of the concepts used changes. The first of these transitions is from symbolic to intangible thoughts and words. It marks the move from direct management to the first level of general management.

The second transition marks the highest level of work in formal organizations. It is the change from intangible to universal thoughts and words used by the truly creative business leaders who anticipate underlying trends and changes to form new businesses that shape global organizations and the communities they serve. It involves creating new markets, seeing new ways of working and devising new products and services that extend human capacity. Information technology is currently being shaped by this kind of genius and the results are radically changing our view of the world and the nature of work.

Human organization structures are shaped by levels of work irrespective of any superimposed formal boundaries. So, in establishing structures and encouraging them to develop, managers responsible for construction need to ensure that the people involved are competent to work at the appropriate level.

Team roles

There is a further factor of importance to managers responsible for making decisions about the people needed to undertake construction work. This is that people have skills and knowledge relevant to distinct team roles. Belbin (1993) provides a classification that has been used by several leading firms in the UK construction industry in selecting individuals to form balanced teams. The aim is to include people who, between them, are competent in all the roles required for effective teams. Belbin's classification is supported by tests aimed at ensuring that within any given team, there is at least one person able to play each of the following eight roles.

The *Coordinator* is the natural chairperson of the team. This person is good at clarifying goals and ensuring that the team agrees on priorities and reaches decisions. The *Shaper* presses for action, finds ways around obstacles and drives the team to action. The *Plant* is the creative source of original ideas and solutions to difficult problems. The *Monitor-Evaluator* carefully dissects ideas that the team are considering, weighs up the options and identifies problems. The *Resource-Investigator* is the extrovert, enthusiastic, inspiring communicator who develops the external links that bring new contacts, ideas and developments into the team. The *Implementor* is the practical organizer who turns ideas into manageable tasks and then schedules and plans them. The *Team Worker* holds the team together by supporting others, by listening, encouraging and understanding. Finally the *Finisher* checks details, worries about schedules and chases others with a sense of urgency that helps the team meet its deadlines.

Belbin's research suggests that all eight roles are needed to form effective teams. One person may play several roles but all need to be provided by someone if teams are to be effective and achieve their best work. Subsequent research has identified a ninth role, that of the *Specialist*. This recognizes the fundamental need for technological competence if teams are to be effective in adding value for customers.

Competent people

So in summary, the first principle of effective organization structures is that the people involved should be competent. This means they are technically competent, multi-skilled, able to work effectively at the level of work required by their task and are competent at one or more team roles. Achieving this may require careful selection procedures, education, training, mentoring and re-training.

Talented individuals will be fulfilled by working on higher levels of work. This does not mean moving up through a management hierarchy. It means tackling long-term issues involving many complex and uncertain factors. It means balancing current activities with future requirements. It means scanning the world-wide environment for developments that may influence current projects. It means working strategically on the basis of values, ethics and cultures agreed within the teams, larger organizations and communities of which the talented individuals are a part. It follows that the second principle of structure deals with these relationships between people.

Relationships

Construction organizations are shaped by the relationships people form and find useful. The significance of this is that organization structures are formed, not by senior managers drawing organization charts but by people communicating. Senior managers influence the patterns of communications by the way they allocate work, decide its location and distribute resources. They can, for example, decide to set up information technology networks that connect some work units and exclude others. However, the more these decisions reflect and reinforce the naturally occurring patterns of communication, the more effective they will be.

As people communicate regularly, private languages develop which allow much routine communication to become virtually implicit so that effort can be concentrated on important matters. There are few misunderstandings and communication becomes reliable in that intended meanings are understood. Shared cultures develop so that individual decisions take account of other peoples' interests. All this takes time but the rewards can be remarkably high levels of efficiency as the confidence that comes from strong relationships frees up nervous energy and creativity.

Managers hope this efficiency will be devoted to achieving agreed work objectives. This requires first, that work objectives are accepted; second, a system of rewards that reinforce the objectives; and third, feedback on the performances being achieved. Managers responsible for construction should give attention to these requirements because they determine the focus and shape of the structures that emerge.

Teams

The first level of organization above individuals is commonly called the team. Teams are important in managing construction, for example Bennett (1991) regards the work of teams as the essential unit of planning for construction project managers.

Many different kinds of team are needed to undertake construction. Some undertake direct construction work that creates an element of the end product, others are responsible for whole projects, while others have a strategic responsibility for a series of projects.

In all these cases the team's objectives may be defined in terms of the required outputs. This means either specifying the products and services to be produced or the performance that must be provided by the outputs. Alternatively, the team's objectives can be defined in terms of specified inputs that are to be devoted to the work.

The objectives, whether specified in terms of outputs or inputs, determine the technological competencies required. The more a team has experience of the required technologies, the more effective it is likely to be. So a team required to find a new way of roofing over an existing grandstand at a major sporting arena may bring together architects, structural engineers, cladding and roofing specialists and construction managers experienced at working together in innovative design. They may decide to bring a tent designer into the established team to provide radically new ideas. The new member will need to be inducted into the team carefully and time and resources will have to be devoted to learning how the expanded team can work together. Given all of this, it is realistic to expect the team to find an exciting new answer and put it into effect competently.

A different situation exists if the required work is, for example, to fix the external skin of a standard warehouse using a well developed cladding system. In this case the team should consist of cladding fixers fully trained and experienced in the specific work. Their work should be planned on the basis of tough performance targets, set in the confidence that the team will achieve them. The team will work virtually instinctively, carrying out well practised activities, using well developed plant and equipment to distribute the cladding panels around the site, lift them into position and fix them. They are very likely to complete the work on time and budget, and zero defects is a realistic expectation.

These various kinds of technological competence, carefully matched to the agreed objectives, provide the starting point in selecting teams to undertake construction work. As with individuals, teams have different levels of responsibility based on differences in their ability to handle abstract concepts and deal with long gaps in time between decisions and outcomes. For example, basic work teams deal with direct physical work for which outcomes are immediately apparent; they can see the results of their work as they leave the drawing office, factory or construction site at the end of each day. Project teams deal with more abstract, generalized ideas for which outcomes become fully apparent only after the project is complete. For example, a construction manager makes decisions about the interactions between basic activities in terms of budgets and programmes; the effects of setting tough targets on costs, progress and defects only become apparent months later when the work is finished and all the claims are dealt with. Strategic teams often deal with even more abstract, generalized ideas and longer time scales. They may, for example, decide to use partnering for their own work and ask project and basic work teams to do the same. Bennett and Jayes' (1998) case studies suggest that the effects will not become clearly apparent for at least three or four years.

Teams asked to undertake basic work, project and strategic responsibilities need to be formed of individuals competent at the appropriate level of work.

Teams at all levels should include individuals who, between them, are able to play all of Belbin's eight team roles if they are to work effectively, but the emphasis changes as work becomes concerned with more abstract and longer-term issues. In general terms, low-level work relies heavily on the *Implementor, Team Worker* and *Finisher* roles, whereas higher-level work puts more emphasis on the *Plant, Monitor-Evaluator* and *Resource-Investigator* roles.

Relationships inside the team should be based on the principles of effective relationships described earlier. That is, there should be clear objectives, effective two-way communication, feedback and rewards that reinforce the objectives for individuals and the team as a whole.

Continuity

Continuity is very important for developing strong relationships and therefore team efficiency. Teams that enjoy continuity learn how to produce the continuous improvements in performance that characterize best practice. This is because individual skills and knowledge improve and relationships between the individual members become stronger. Gradually each member learns where they fit into the team and what they can rely on from the others. They learn to compensate for individual weaknesses and to use individual strengths to the best effect. They develop an internal language based on shared experiences and common understandings of terminology. Common cultures and values guide their decisions and they work together almost instinctively. The result is very high levels of efficiency and competence.

Generally this means that teams should be kept together as long as their broad objectives and the circumstances influencing the achievement of those objectives remain substantially unchanged. The most robust basis for continuity in construction teams is consistent patterns of technologies. This provides many more opportunities for continuity than the traditional emphasis on the function of end products because there are relatively few effective clusters of technologies compared with the great multitude of different facilities produced by the construction industry.

A good classification of technology clusters is given by Gray (1996) who suggests the following technology clusters for building projects:

- Substructure,
- Structure,
- Envelope,
- Service cores, risers and main plant,
- Entrance, finishes and vertical circulation spaces,
- Finishes to horizontal spaces.

For each of these technology clusters there are few options that are sufficiently different to require unique combinations of skills and knowledge. This makes it relatively easy for teams responsible for technology clusters to achieve continuity in their work. So it is sensible to invest time and money in developing better ways of working and improving their technologies. Similarly project and strategic teams using consistent patterns of technologies to produce the products and services needed by broad categories of customers can afford to invest in developing better answers.

Limits to continuity

There are, however, limits to continuity because teams develop and become capable of dealing with longer-term and more complex issues. As they master the problems thrown up by any given level of work, much of what they do becomes routine and they start to become aware of new and more challenging levels. This is the essential restlessness of human nature. For many talented people, once they have solved a problem it ceases to be interesting and soon they look for new issues to tackle. So as the answers to old problems become second nature and can be produced virtually automatically, well motivated teams look for new challenges.

As described earlier in this chapter, each level of work requires a different way of working. Each level builds on and indeed depends on the skills and knowledge learnt at lower levels. It is natural for teams to develop and grow so, unless their work changes with them, their original level of work becomes boring and they become frustrated. Faced with this situation, the most talented or ambitious individuals often decide to leave a successful team for a higher-level one. They feel they have outgrown the rest of the team and move on for the sake of their own career. This has a disruptive effect on the team. It is extremely difficult to replace key members of teams with someone who provides the same combination of technical competencies and team roles and is able to work at the right level. Even if this is achieved reasonably well, the new member has to learn the language, culture, values and work methods that have become second nature to the rest of the team. In most cases the situation is worse than this because the ambitious member who leaves is replaced by a person with a lower level of skills and knowledge. Where this happens frequently, teams may remain competent at a consistent level of work but at the cost of constantly having to learn how to work with new members. As a result they fail to achieve the high levels of efficiency that real continuity provides.

These risks make it particularly important to provide opportunities for individuals to develop and grow inside established teams. Otherwise ambitious individuals have no incentive to remain and some at least will choose to leave causing the kind of disruptions and inefficiencies described above. In practice the development of individuals and their orderly movement into and out of teams needs to be planned. The careful introduction of new members can transform a team by bringing in skills and knowledge previously missing. There will inevitably be an initial period of adjustment but this can be very creative in opening new possibilities for the whole team. Indeed, when teams get stale, it is often a good idea to introduce a new member. This can revitalize a team and help it to make an effective new start.

So continuity for teams provides many benefits but needs to be planned flexibly to ensure that they continue to improve their performance.

Continuity of course depends on teams being provided with a steady stream of work. The broader their skills and knowledge, the easier this is to achieve. Never-the-less the work needs to match their growing strengths and so gradually involves higher levels of work.

Low levels of work still need to be done and good teams learn how to deal with the lowest levels of their own work in ways that avoid its frustrations. They find machines to do some of it, they subcontract other parts, or they change the products and services they produce to eliminate it altogether. As their work becomes concentrated on higher-level work, they must learn how to apply their growing skills and knowledge to ever wider ranges of situations in order to maintain a steady flow of work. This accelerates the development of new skills and knowledge and the need to move to higher levels of work. Eventually new teams are needed to undertake work the original team is no longer prepared to do.

Therefore continuity is not a simple, unchanging situation. It requires a dynamic and ever-changing balance between the level of work undertaken by well established teams and career paths for individuals that motivate and reward them.

This dynamic balance is shaped by changes in construction technology and markets. It is influenced by teams developing and growing as they change their members and their competencies. New teams are formed to undertake low-level work as established teams move on or disband. Throughout all this change, continuity for individual teams is a potent force for improvement and development. So structures that reinforce continuity are more efficient than those dominated by change and discontinuities. Yet change is an inevitable feature of the human condition and of the environments in which construction operates. It is reasonably certain that some degree of change is necessary to trigger new ideas. Hence, effective structures provide a dynamic balance between continuity and change. We shall see how the overall structure of the construction industry can provide this and so deliver both efficiency and innovation. Before we turn to this overall structure, we need to consider the nature of team interactions.

Team interactions

The first two principles of structure are that effective organizations require competent people linked by strong relationships shaped by the needs of the work. This produces teams competent to undertake specific construction work. However, the structures which undertake construction are shaped by interactions between teams. There are three bases for team interactions: technology, territory and commercial.

Technology

Technology is the main factor at work on individual projects. Interactions are determined by the need to bring together teams able to produce a design, establish a budget and programme, construct foundations and underground services, construct load-bearing structures and so on. Distinct kinds of project team emerge from recurring patterns of these technologies. The most efficient develop consistent forms of interaction based on the experience of working together over time.

Technology is the driving force in supply chains that turn basic raw materials into elements of constructed facilities. The teams that assemble elements on site are only one part, often a very small part, of complete supply chains. Many of the most remarkable improvements in productivity in industry generally have come from investing in the development of supply chains. Much of this improvement comes from strengthening the interactions between teams throughout supply chains.

Likewise, teams involved with work that supports similar functions should interact. Thus those involved in housing, hospitals, education facilities or defence installations should work together to produce better facilities and support services. Trade and professional organizations often arrange their activities around common types of buildings or other constructed facilities. The best encourage networks that link activities across the separate disciplines. The initial basis for these interactions is the common technologies used in the particular type of facility and the need to understand how these can best be accommodated. As the best answers are established, the basis for the interactions becomes more fundamentally grounded in the resulting patterns of construction technologies.

Teams interacting on the basis of technological interdependencies benefit from cooperating in the development of their methods of working and their products and services. Ideally they will undertake this development work independently of specific projects in cooperation with leading customers, designers and managers responsible for distinct types of projects. At their best, these networks aim at building the market for their particular technologies by delivering better quality and value at lower costs. In taking these initiatives it is of course sensible to make sure that they also provide higher profits.

Territory

The second basic factor shaping interactions is territory. Teams working in one geographical area should interact with the aim of developing an effective local industry able to contribute positively to the community of which it forms part. This gives rise to rich patterns of interactions with many local and regional organizations, including those with political, financial, social and technical interests.

Teams involved in technology clusters on individual projects share common work spaces. This territorial factor is important but relatively short term. It adds little to the technological interdependencies. However, it needs to be taken into account in understanding the interactions within project teams.

Commercial

Teams also interact because they are linked commercially. The teams may be part of one firm or be formed jointly by a group of firms cooperating on a programme of work. They may have responsibilities for different levels of work that contribute to a complex, widely distributed initiative. Global organizations usually need to work with small, local organizations in different countries when they are tackling international issues. These often include construction teams who then find themselves in networks that link regional, national, international and global teams contributing to one inter-related enterprise. Whatever the particular circumstances, commercial considerations are a factor in determining the interactions of most construction teams.

Practical implications for construction

These three bases give rise to richly interconnected networks. As with relationships between people, interactions between teams develop and grow strong with repeated use. The interactions should be guided by the same general principles as those between people inside teams. In other words, effective interactions between teams require objectives, communication, feedback and rewards. The way they are provided in practice is largely determined by the overall type of work.

The new paradigm identifies two broad types of work that shape construction's overall structures. These are mainstream and new-stream work, each of which gives rise to structures that comprise teams with different competencies and which interact differently. The principles that shape these structures are the same, but the type of work causes marked differences in their practical application.

Mainstream work produces products and services that meet the needs of the broad majority of customers. Firms involved in mainstream work should provide reliable good value on simple, straight-forward terms that satisfy the needs of specific categories of customers. Different firms will concentrate on different market sectors so that all significant mainstream needs can be met. The primary requirement in mainstream work is efficiency so the majority of construction needs can be met at prices customers can afford and yet at which construction firms earn fair profits.

New-stream work produces new designs, new technologies and new ways of working. Firms involved in new-stream work need to be innovative. They work with individual customers to develop new kinds of facilities. They work with their suppliers to devise new forms of construction. They bring teams together to search for the best possible answers to challenging opportunities and problems. Their work is more expensive and less predictable than that of mainstream firms. Never-the-less, the best new-stream firms give customers security by being clear and open about risks and working to carefully defined budgets and programmes.

The construction industry should differentiate between mainstream and new-stream because excellent performance comes only from organizations that concentrate on a consistent type of work.

A particular problem that has to be dealt with by the overall structure of the industry is that the focus which provides efficient mainstream work can be carried too far so that organizations become inflexible and unable to respond to change. It is necessary for mainstream work to develop in response to changes in markets and technologies. This means that in addition to relentlessly improving their current methods, products and services, mainstream organizations should search out problems and opportunities and devote resources to finding good answers. This may be done internally by setting up a task force or project team, which can usefully include external experts, some of whom may be drawn from new-stream organizations. Alternatively, an external organization specializing in new-stream work may be commissioned to find a radically new answer. Once a robust answer is found, it should be introduced into mainstream work by the team that developed the new idea working jointly with a mainstream project team.

These interactions between mainstream and new-stream enable the industry as a whole to deliver efficiency and innovation. It means that mainstream and new-stream organizations are distinct but linked. The following descriptions of these two kinds of structures include discussion of the financial arrangements used in best practice because these necessarily reflect and influence the relationships in these organization structures.

Mainstream structures

Mainstream work depends on continuity so that teams can develop strong internal relationships and strong interactions with other teams. These strengths come from continuity in the patterns and sequences of work so that teams at all levels have similar work and consistent interactions with other teams over many projects and long periods of time.

Continuity results either from marketing a range of products and services or working for a major customer with a regular programme of similar construction work.

The approach adopted by a number of leading customers in the UK provides strong evidence of the benefits of continuity. Bennett and Jayes (1998) describe substantial improvements in the costs and completion times of new building projects undertaken for customers with regular programmes of similar work. The results produced by this second-generation partnering, given in *Table 1.2*, show that cost reductions of 40 per cent and time reductions of 50 per cent can be achieved. In all the cases described, the customers took the initiative in setting up teams to look for systematic, continuous improvements independently of individual projects. These customers tend to be in businesses exposed to international competition which forces them to question all their costs, including the costs of their constructed facilities. As a key part of their overall business strategies they have provided continuity for consultants, contractors and specialists in return for cooperation in achieving continuous improvements in their joint performance. In most cases the focus initially is on cost reduction. However, once the benefits of continuity become apparent, customers aim at broader improvements, including better designs, faster completions and fewer defects.

These important developments have given parts of the UK construction industry the opportunity to break free from their traditional project-based approach to think and plan long term. In the main this is tied to the work of specific major customers but it has given the construction firms involved the chance to learn how to steadily improve their products and services and so improve the value for money they can offer customers. This gives them a wonderful marketing opportunity based on a track record of delivering reliable value for customers. In the long term, marketing this kind of soundly based value for money is the only robust way of providing the continuity on which mainstream work depends.

Diverse customers

The construction industry traditionally has not used marketing to provide continuity because customers' needs are seen as too diverse to be met by standardized ranges of products and services. It is true that customers are diverse in terms of the use to which they put their constructed facilities. Also they are different in terms of size and experience of commissioning construction work.

This overall diversity is complicated by many important customers being diverse internally. They often have many divisions and levels involved in establishing the business case for a construction project and with an interest in how it is managed. The internal and external diversity causes the construction industry to see most projects as unique. This perception is reinforced by a real problem, which is that all too often the people involved in preparing the business cases for new construction are mainly concerned with financial matters and technical aspects of the customer's business. This means they have little or no property or construction industry expertise. As a result the business case focuses on the customer's specific requirements which means the resulting construction work is described in terms of an individual, new-stream project. By the time people with construction expertise become involved, opportunities to use established, efficient answers have been closed off. This is particularly the case with one-off customers who are encouraged by creative designers to look for individual answers.

Bennett and Jayes' (1998) case studies show that the pressure to adopt new-stream answers is being resisted by experienced customers. In order to get what they want in terms of efficiency and reliability, they have become closely involved in the work of the construction firms who undertake their work. Customers have been forced to get involved because the construction industry has failed to take the initiative in developing mainstream answers they can use. These customers have insisted on what is essentially a mainstream approach of using established standard answers and then working at continuously improving them.

It is a relatively small step from mainstream answers developed for individual major customers to generic mainstream answers that suit broad categories of customers. The key is to recognize that customers have essentially three interests that construction needs to satisfy. First there are financial interests; the customer has to be able to fund new construction and the subsequent maintenance of the new facility. Second, there are the interests of the people who will use the new facility. A new facility that helps users be happier, more productive and successful and is generally enjoyed by them provides a very powerful marketing message. Third, the interests of customers' facilities managers need to be satisfied. They want facilities that are easy to run and maintain, help them deal with changes as they occur, do not spring surprises and generally support their role in the senior management of their organization's business.

These diverse interests mean that many people within a large customer's organization have an interest and an influence on construction work. The industry traditionally has avoided getting involved in these internal issues with the unfortunate result that construction firms are seen by many customers merely as minor players in their overall business strategies.

A common traditional approach is for the industry to advise customers to appoint one internal person to provide a single point of contact for each new project. This is intended to protect construction from the customer's internal politics by providing a clear brief and a consistent way of dealing with problems. It is unrealistic to expect any one person given responsibility for a one-off project of any size or complexity to provide this kind of certainty. Internal problems will arise and changes will be necessary. The construction industry's traditional response of making a claim simply alienates customers and makes it more difficult for them to trust construction firms in discussions about future construction projects.

Dealing with diversity

Construction organizations need to invest in understanding their customers' financial, use and facilities management interests to provide the basis for sustainable continuity in their work. This broad understanding is essential preparation for them to develop ranges of mainstream products that give customers real choices in how each of their key interests can be met. Once ranges of mainstream products have been developed, they need to be marketed by specialists in each of customers' three key interests.

The marketing specialists should include experts in financial engineering able to deal with the interests and concerns of every potential customer, including those concerns raised by financial departments in major international companies right through to the simpler needs of individuals wanting new construction. So the financial specialists will understand the costs and methods of financing construction. Equally they will have expertise in the income and other financial benefits that can flow from new construction. This deep knowledge gives them a basis for devising a range of financial packages that make it easy for most potential customers to afford new construction work.

The second kind of marketing specialists are experts in the use of constructed facilities. They should be able to show whoever will use a proposed new facility how it could help them. The specialists need to provide clear descriptions of the options available. Effective ways of doing this are to visit existing facilities or to use computer-generated visual displays which, at their best, become virtual reality. Also workshops that bring together representatives of the users, creative designers and researchers into various aspects of the particular kind of facility help build a deeper understanding of how the new facility can be used. An important part of understanding how facilities will be used is to identify what Carlisle and Parker (1989) call the ultimate customers. These are the people who make choices between buying the customer's products or those of a competitor. Understanding the ultimate customers' motivations and decisions will help the construction organization to understand how a new facility can best help the customer's business.

The third group of marketing specialists are experts in facilities management. They will help the customer's staff responsible for running and maintaining the new facility understand how it should be operated to get the best performance at the lowest costs. They will have maintenance strategies that cause the least disruption to users and minimize overall life-cycle costs. They will offer customers ranges of after-care services at predetermined annual costs. The overall effect will be to make the work of facilities managers easier and more effective.

Potential customers will make best use of construction organizations' broad and well informed marketing competence if they appoint an internal project manager. This role is to act as a gatekeeper, providing access for construction organizations and their marketing specialists to influential teams within the customer's own organization. The gatekeeping role is needed because, as discussed above, many different teams have a legitimate interest in shaping the customer's objectives but it helps to have a single point of responsibility and contact with construction organizations. This is true even when the customer knows which construction organization it wants to undertake the new project and simply appoints them. It is even more the case where the customer needs to make a selection from amongst a number of competing construction organizations.

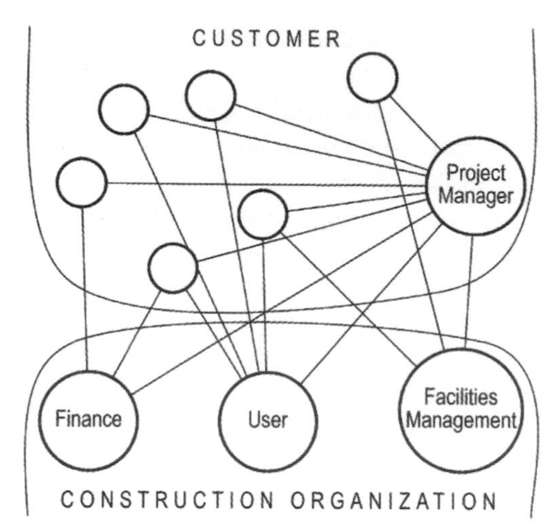

Figure 5.2: Interactions between customer and mainstream construction organization

The project manager's job, as *Figure 5.2* suggests, is to ensure that all the internal interests, including those that potentially conflict, are identified and taken into account. He or she also acts as an internal champion for the project when problems or concerns arise internally. This can only be done effectively when the project manager works by facilitating interactions between the internal and external teams involved. In no sense should the project manager act as a barrier, or take unilateral decisions in areas that impact the interests of colleagues.

The structure of interactions found to be necessary between customers and the construction industry should determine the choice between mainstream and new-stream construction. When the required construction work can be agreed quickly and simply by reference to an established answer, the work is mainstream. The better the construction industry becomes at marketing well developed mainstream answers, the more often will customers decide mainstream provides the right answer. This is how continuity can be built up by marketing.

Work should be treated as new-stream only where the interests involved give rise to rich and complex networks of interactions between the customer's organization and external construction teams. This should be a minority of cases because the long-term health of construction depends on developing mainstream answers that suit most customers.

Forming mainstream organizations

Mainstream work is characterized by well established roles that give workers at all levels a robust basis for efficiency. Individuals undertake work they are well used to doing, the relationships inside teams have grown into strong bonds and interactions between teams use well established procedures.

Individuals in effective mainstream organizations have a broad competence and are multi-skilled so they can communicate, work in a team, reflect on their own situation and continuously improve their own performance. This requires well established standards and procedures that encourage them to look for better answers. Also, mainstream workers need training in core skills such as process analysis, planning work, problem solving, communication and performance measurement. They often need help in moving beyond the traditional boundaries of their own professional or craft discipline so they can deal with problems and new ideas as they arise. This flexibility helps insulate mainstream work from external interference so workers can concentrate on working efficiently.

This kind of robust efficiency does not come easily. It requires new entrants to the construction industry to be educated in the core competencies of design, engineering and management and to have a general understanding of mainstream and new-stream work. Then they need training in a range of work roles until they can decide where they will pursue their careers. This means organizations should allow people either to concentrate on improved efficiency in some aspect of mainstream work or to pursue new-stream ideas. Providing these choices may well require mainstream and new-stream firms to cooperate in training.

The resulting competent people work for various kinds of firms. Each will be competent in some of the key skills and knowledge needed for mainstream organizations to be able to offer potential customers a complete package of products and services. Consequently, mainstream work usually requires several firms to cooperate.

In deciding which firms to bring together to form a mainstream organization, it is sensible as far as possible to gather clusters of firms that like working together. This issue is discussed under the membership pillar in Chapter 4, as is the decision on whether the organization should include just one or more than one firm of each type. The key is sustaining a reasonable level of work for all the firms in order to provide sufficient continuity to justify investing long term in continuous improvement.

Having selected the firms to form a mainstream organization, it is important to ensure that cooperation is a reality at all levels of interaction. Narrow contractual attitudes between firms at one level break down cooperative behaviour elsewhere. To help stop this happening, best practice gives one person in each firm overall responsibility for ensuring that their firm acts cooperatively at all levels in dealing with the other firms. Giving one individual the authority to make key decisions helps ensure that their firm acts cooperatively in its external relations. Where the cooperating firms have formally adopted the disciplines of partnering, these individuals are commonly called partnering champions.

Senior managers should encourage teams at all levels to build internal and external structures that facilitate trust and provide many opportunities for communication and feedback. In building these bases for cooperation, it is important to recognize that middle managers within organizations probably have to take the biggest risks when cooperating, especially if cooperative working is not fully supported at senior management level. Middle managers are taught to be defensive, both by their training and experience of traditional methods, but joint work with other firms requires them to work in a cooperative manner. This is easier to achieve when firms maintain consistency in the people who interact at all levels but especially at middle levels.

It is very helpful in building cooperative behaviours if firms invest time and money in openly supporting the aspirations of the other firms they are working with. This often means taking risks in areas of common ground or joint interests. People are unlikely to take risks for the benefit of another company if they cannot see some clear payoff for themselves. Trust has to be based on two-way benefits. Problems arise, for example, if a buyer is not working consistently at reducing the variety of components demanded and simply treats suppliers as free warehouses. It is important that main contractors do not crush smaller specialist contractors under bureaucracy or unreasonable demands. People will not be trusted if they ignore the interests of smaller firms. Cooperative behaviour has to be two way.

Workshops help to build and reinforce cooperative behaviour. They encourage firms to be open about their own business interests for the good of their joint work. It is very motivating when the people involved in a workshop feel confident enough to be totally honest about their firm's interests, including the need to earn fair profits. Workshops are very effective at encouraging all the key interests to be openly discussed by teams before mutual objectives are agreed. Then, provided all members of the team are given every opportunity to concentrate on doing their best work, it is likely that the agreed objectives will be met.

Mainstream work

In practice, mainstream work has adopted several distinct structures. These mainly result from differences in the focus of the work. Some concentrate on continuously improving the product, that is standardizing the building or other facility, whilst others focus on standardizing and continuously improving the design, manufacturing and construction processes. As a result there is a range of approaches that, viewed in total, show that mainstream thinking can be applied to many different types of facility. The following examples, which all relate to buildings, are described more fully in Bennett *et al.* (1996).

McDonald's use factory-made modules to produce fast food outlets that can be assembled on site in just a few days. They have worked with manufacturing and construction firms to develop and continuously improve a range of eight standard buildings that meet all their needs for stand-alone fast food outlets. Once all the necessary approvals are available, the current version of the selected standard building can be produced very quickly, thus allowing the income stream from a new outlet to begin much earlier than with traditional methods. The buildings match the performance levels of traditionally built outlets, currently at the same cost, however the cost is relatively inflation-proof because of the potential for continuous improvement in the controlled environments of factories. This highly integrated approach has considerable potential to provide ever better value for money and continuously improve performance.

A different approach has been adopted by Waitrose for their new supermarkets. They have developed three standardized design processes, each catering for a distinct market. The design processes are held in computer systems that automatically produce complete designs tailored to individual sites. Once a new site is available, and the managers who will run the supermarket have decided on the mix of goods they will sell, the most appropriate of the three design processes is selected. It is applied straightforwardly taking account of the specific characteristics of the particular project. This allows individual projects to be undertaken efficiently by concentrating on realizing a predetermined design. Good ideas that emerge during the project are used, not on that project, but are considered carefully, properly developed and added to the standardized design processes for use on future projects. The design processes are also subject to continuous improvement based on feedback from existing supermarkets and R&D work.

Similarly Whitbread have developed generic designs for the various types of buildings they need for their leisure business. Sainsbury's also have a standardized approach to the main elements of their supermarkets. Esso have standardized the design of their service stations so that the only important variables are the anticipated volume of business and the configuration of the site. This approach has been further developed with the use of factory-produced modules for major elements of their service stations. Gazeley Properties have a highly consistent approach to the design and construction of large distribution warehouses. At present this is implicit in their use of a small range of consultants and contractors who are very experienced in the specific technologies used. The widely reported approach adopted by Stanhope for their development of world-class office buildings at Broadgate in the 1980s provided for continuous improvement from phase to phase within one mega-project. Similarly they arranged for carefully selected designers, managers and specialists to produce highly creative industrial buildings at Stockley Park, near Heathrow, London by ensuring that each new building applied the lessons learnt from earlier ones.

Many other examples exist and they all fit within the range of approaches illustrated in *Figure 5.3*. The range begins at one extreme with approaches that are a simple extension of traditional methods, providing one-off designs of very similar buildings produced by a consistent group of consultants and contractors. They consciously develop a limited choice of technical solutions from project to project or, on a very large project, from phase to phase.

The second approach is to consistently develop a design from project to project. In this approach the

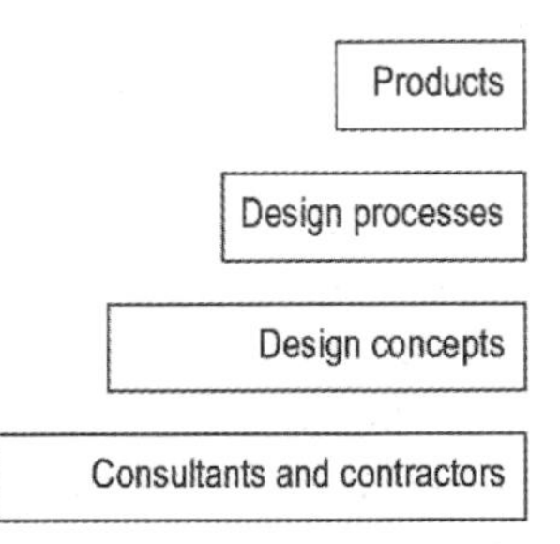

Figure 5.3: The hierarchy of standardization in the continuous improvement of mainstream work

previous design is used as the starting point for each new project. It is altered as little as possible to meet the new customer's needs.

The third approach is to develop a small range of standard design processes that are continuously developed, with each new building using the current version of the most appropriate standard.

The fourth approach is the closest to full mainstream thinking in current UK practice. It is based on producing a small range of standard buildings that are continuously developed, with each new building using the current version of the most appropriate standard.

The emergence of these approaches based on the systematic, continuous improvement of construction products and processes is a major development in the work of the construction industry. It is at present almost entirely customer-led. Its significance as yet is scarcely recognized by the industry, still less taken into account in its internal methods. Once this hurdle of understanding has been overcome, the key change will be for the industry to use systematic feedback to provide a steady stream of improvements that meet customers' real needs. There is no shortage of ideas and creativity, as evidenced by the industry's traditions of experimenting with many new styles, technologies, materials and methods. The weakness of this traditional creativity is that it has no overall direction and so innovations are random. The range of more systematic approaches now emerging shows that the industry's talents can be focused onto mainstream work. The key is to use feedback to provide steady, continuous improvements in performance.

Structure for feedback

Feedback is crucial to achieving control and then continuous improvements in performance. Deciding on the specific feedback to be collected and how it is to be used are crucial decisions for the strategic teams. The basic elements of any feedback system are that a clear target is set, progress towards the target is measured and exceptions are identified. This is implicitly understood in the many cost and time control systems used throughout the construction industry. There is great merit in making this understanding explicit and being tough about applying the disciplines needed for feedback-driven control systems to become virtually automatic.

A starting point is for teams at all levels to agree how improvements in performance are to be measured. This is complicated when several firms are involved. A good place to start is with measurements that are routinely produced within the individual firms as part of their management control systems. The ways in which these are produced and used should be reviewed by a joint team; the aim being to identify what is important and can be measured accurately and efficiently. It is helpful to consider both objective and subjective measures. The judgement of experienced professionals, especially where several people make independent judgements and then debate them, is an entirely valid basis for measuring the performance of teams. This is especially so where subjective judgements are combined with well defined objective measurements to provide a broad evaluation. All these issues need to be considered and out of the ensuing discussions teams should identify a small number of clearly understood measurements that reflect the improvements in performance they want to make.

Ensuring that feedback systems are in place and working effectively in all the separate teams is often a key role of the partnering champions. They need to ensure that reliable feedback data from individual projects are collected and systematically analysed to identify trends in overall performance. The methods used in collecting these data should be monitored to ensure that they are accurate and comprehensive. The results should be discussed regularly by all the teams involved. Their conclusions should be used to maintain work within quality, time and cost targets. The data should also feed into reviews of the standards and procedures used by project teams so that improvement is automatically built into normal working methods. The objectives set for individual projects should be guided by feedback so the targets are both achievable and challenging.

Feedback should include information about products and services. It is good practice to carry out surveys at regular intervals during the first two years of occupancy of a new building to capture the experience of those who use and run new facilities. Thereafter annual surveys help construction organizations develop a robust understanding of their own products and services. They also provide early warnings about changes in customers' needs. Records of complaints from customers and how they were dealt with are another important source of feedback. In much the same way it is important to collect feedback from firms who make up the organization's supply chains. This should include information about their experience of working as part of the overall organization as well as about problems they have encountered and the outcomes. Feedback in all these various forms is the driving force which ensures that mainstream organizations are delivering continuous improvements in their overall performance. It also provides warnings of the need for a step change.

The final important point is that feedback should never be used as a basis for blame or punishment. Its purpose is continuous improvement and so should always be used positively.

Organization structures

Figure 5.4 illustrates a mainstream organization structure. It shows the main parts that exist at one point in time and one possible pattern of interactions between them. At other times the organization is likely to include other parts and other interactions may become important. The structure is a flexible network formed by setting up

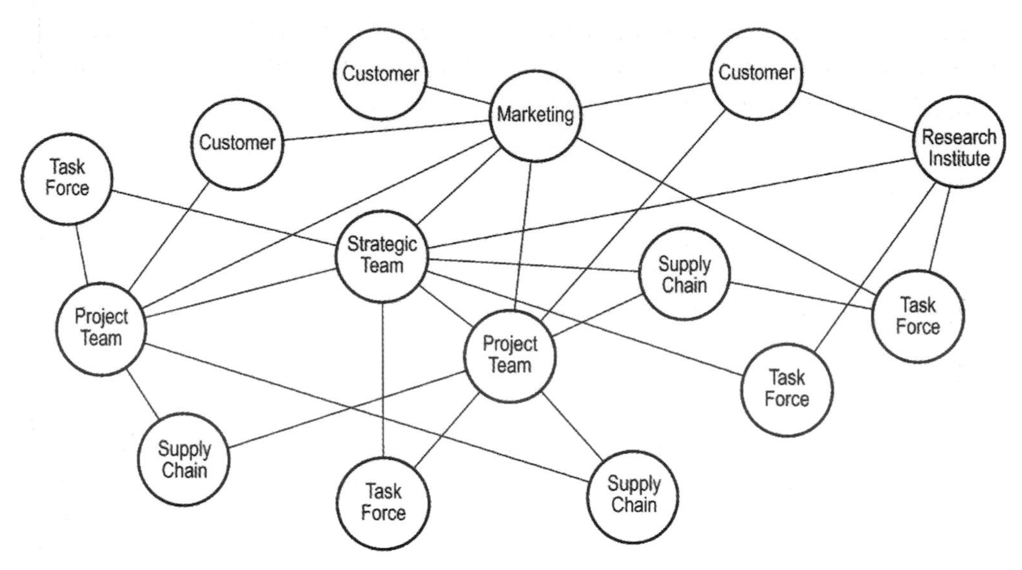

Figure 5.4: Main parts of mainstream structures

teams with broad responsibilities and empowering them to communicate with whoever can help their work. In addition to the interactions shown in *Figure 5.4*, each of the teams will form external links to help them undertake their work.

Strategic team

The main parts of the mainstream organization include a strategic team formed of representatives of all the firms, or all the divisions of one large firm, that together form the mainstream organization. Its role is to ensure that the overall organization has ranges of well developed products and services that meet the needs of specific categories of customers. These will be embodied in formal standards and procedures that guide the work of the organization. The strategic team should ensure that the products and services are steadily and continuously improved on the basis of systematic feedback. They also need to ensure that their workload is sufficient to provide continuity for all the teams in the overall organization.

Marketing

The organization includes marketing specialists who are centrally involved in sustaining the workload. They need reliable information about the organization's products and services to guide discussions with potential customers and deal with all their queries and concerns about financial, functional use and facilities management issues. More importantly, the marketing specialists need sufficient information to provide a robust basis for the offers and promises made in negotiating agreements with customers.

Customers

Experienced, repeat customers form another key part of mainstream organizations. They will normally interact with other teams on the basis of simple and straightforward agreements about new projects. These may be initiated by the mainstream organization's marketing specialists identifying a new opportunity for their standard products and services to help the customer's business. Equally the agreements may emerge as part of the answer to a problem or new opportunity being tackled by the customer. Whatever their origins, the agreements will often be concluded quickly and include little more than decisions about which standard building to construct on a given new site, how the peculiarities of the site will be handled, the way the interests of neighbours are to be dealt with, the support services to be provided, the completion date and the price. There will be confidence in both parties that all significant interests have been met and that the agreement will be fully implemented.

An important characteristic of mainstream work for experienced, repeat customers is that interactions between them and other parts of the construction organization are strong and well developed because both organizations are working in predictable ways with people they know and trust.

One-off or occasional customers, or those the organization has not worked with before, are also important parts of the overall organization but they need to be dealt with differently. The structure of interactions has to be formed first, before all the details of the agreement can be negotiated. In essence the marketing specialists work through a decision tree which helps them understand how the customer's organization operates, what is its culture and style, and helps them to understand all the interests that influence the possible new construction work. An integral part of the decision tree will be helping the customer's organization understand how new construction can help their business. Joint workshops are often used to develop a common understanding of the most important issues and to build trust and cooperative attitudes between the key actors. Gradually a structure of interactions will develop capable of negotiating a sensible agreement. Initially it will be tentative in many respects, so the overall arrangements will include a process designed to make joint decisions as and when problems and new issues arise.

It is likely during these early stages that a new customer will discuss his project with three or four mainstream organizations. This competitive element helps reassure the customer that he is getting good value and that no significant options are being ignored. So mainstream organizations need to have structures that are sufficiently flexible to fit the needs and aspirations of most customers. The aim of the initial negotiations is to guide the customer towards some combination of the organization's standard products and processes or, alternatively, clearly identify that a new-stream answer is needed. In the first case the organization will hope to reach an agreement with the new customer and, in the second, advise them to use a new-stream organization. The integrity of providing honest advice, even where this means a mainstream organization recommending a new-stream organization, is essential for the industry to improve its reputation and build a body of loyal customers.

Supply chains

Other important parts of mainstream organizations are efficient and reliable supply chains for all the elements of the organization's products. These are responsible for decisions about development work, agreeing designs, quality standards, prices, call-up periods and methods for dealing with problems. In most cases any one supply chain will consist of several firms, each undertaking a distinct stage in the overall process of turning basic raw materials into a major element of the end product. Similar considerations apply to the services the organization offers to customers. They, too, usually involve a supply chain of firms providing specialized aspects of the services. The strategic team needs to set up teams responsible for each of the product and service supply chains whose job is to ensure that all the links are efficient and reliable, and that the firms involved actively search for continuous improvements. Supply chains are important sources of new ideas because many of the firms involved invest heavily in R&D work and serve wide-ranging markets.

Project teams

Project teams are another important part of mainstream organizations. Their tasks are based on agreements reached by the marketing specialists with the customer. These will normally be to provide the customer with the most appropriate of the organization's standard products and services. Project teams will consist of a small number of experienced designers and managers who put a well established plan into effect to provide everything promised to the customer. At least one member of the project team will have been closely involved in reaching the agreement with the customer to ensure that there is a deep understanding of all the needs and requirements.

Project teams will work with supply chain teams to coordinate the work of the firms responsible for manufacturing and assembly work. They will use comprehensive quality, safety, time and cost control systems to identify problems early. If a problem arises the project team will concentrate all the necessary resources into solving it quickly. They will work with the marketing specialists who negotiated the original agreement to keep the customer informed about progress and then hand over the fully complete product exactly on time. Finally, they will review the project to provide the strategic team with feedback on their performance against measured benchmarks and all the lessons, good ideas and problems encountered.

Task forces

The overall organization is also likely to include task forces set up to develop a better answer for some feature of their products or services, find a better way of working, or tackle a recurring problem. Task forces bring together people experienced in the particular aspect of their organization's work, including external experts. They will be given the time and resources needed to find the best possible answer. They may well work with a project team in testing various ideas. They often commission new research or development work to be undertaken by external organizations. In many cases, the organization will have an internal research institute that is closely involved in the work of task forces. Once a good answer has been tested and shown to provide real benefits it will be added to the organization's standards and procedures to be used on future projects.

Interactions

The various parts of the organization interact in carrying out their particular work. The interactions that prove to be useful will grow strong and, in so doing, will determine the organization's structure. The driving force behind all of this should be feedback from customers, projects, supply chains and task forces. This guides the strategic team in responding to changes in the organization's markets and new developments in technology. It helps all the other teams work in accordance with the overall strategy and to find ways of improving their performance.

Long-term development

Joint development work needs to be organized so the firms involved remain committed to supporting it. It is easy to react to difficult market conditions by cutting back on long-term development work. There are few penalties in the short term but the long-term damage of not continuing is a gradual loss of competitivity.

There are many studies of the UK economy which conclude that these short-term attitudes have been all too prevalent in industry generally. Construction has fallen into this trap; never-the-less there are important examples of successful long-term developments in the UK construction industry. There need to be many more of these but the lessons from the examples that do exist are clear and they are consistent with case studies of long-term development work undertaken elsewhere. They show that best practice includes setting up multi-discipline development teams and ensuring they have the time and resources to find, test and apply new answers. Staff are rotated between the development team and mainstream practice, usually on a three-year cycle. The development team is set clearly defined objectives that are broad enough to encourage creativity and innovation but sufficiently tightly defined to avoid the work drifting into open-ended research.

It is important that there are close links to live projects. A key part of this is the development team being called in to help project teams solve difficult problems. The development team also needs close links to research institutes so they know what is happening at the leading edge of their subjects. The development work should make systematic use of tests and prototypes to ensure that new ideas are practical and effective before they are handed over to project teams. When a new idea is ready to be introduced into mainstream practice, this should be done jointly by the development and project teams. The final key point from best practice is that development work should take account of well organized feedback from live projects to ensure its continuing relevance.

Step changes

Mainstream organizations need to subject their mainstream work to steady continuous improvement. However, there are limits to the improvements that can be achieved by the steady development of any given technology. There inevitably comes a time when the organization must make a step change to new answers. It may be that competitors are developing better products or services, or that customers' needs have changed. Today's global markets make these major changes more frequent and more likely.

It is too late to begin the search for radical new answers when the need for change has become clearly evident. Mainstream organizations that survive long term ensure that when a step change is needed, a number of options already exist.

The creation of potential step changes is organized in several different ways by successful mainstream organizations. Cases exist where a distinct organization is formed to undertake R&D. This may be a separate organization funded through a formal joint venture agreement or may simply be a long-term task force staffed and funded by several firms agreeing to cooperate.

It is important for mainstream organizations to recognize that innovation often emerges from activities at or outside the boundaries of the formal organization. The people involved in such virtually clandestine work are often called skunk groups. They develop ideas to which they are highly committed but to which their firms are unwilling to allocate resources. So they cheat by using time, materials and equipment without authority in order to produce something new. De Geus (1995) describes a study of firms that have stayed in business for a very long time. It shows one important thing they have in common is a tolerance for skunk groups. They know that the high commitment generated from allowing people to work on things that really interest them benefits all their work. More importantly, experience shows that when firms face radical change which threatens their mainstream business, their salvation often comes from a skunk group having developed an idea that provides a way forward.

Best practice mainstream organizations reflect this experience by ensuring that everyone knows it is acceptable to look for potential step changes. The aims should be to ensure that mainstream work is undertaken efficiently through fully integrated structures whilst, at the same time, allowing people the freedom to explore new ideas and potential new opportunities. Then, when customers' needs change or competitors devise better products and services, the organization can respond quickly and effectively to maintain its market share. Ideally of course the organization will lead these changes but that is not always possible. In any case organizations that want to stay in business long term need a range of potential step changes to give them flexibility in addition to their mainstream efficiency.

Financial arrangements

When mainstream organizations are fully established, they work with customers to produce the business case for new construction work. Their role is to suggest interesting ways in which new construction can contribute to the customer's business and assess the costs and risks involved in the answers being considered. They evaluate potential problems caused by restricted access to proposed sites, poor ground conditions, the availability of services, difficult neighbours, restrictions on development and anything else likely to influence what can be constructed or its costs.

When a well considered business case is agreed, they offer the customer a product and associated services for a guaranteed fixed price and completion date that are fully in line with the customer's business case. In addition to a price for the new construction work, they may include prices for facilities management services to run and maintain the new facility. Whatever range of prices are offered, the mainstream construction organization will use independently researched benchmarks to provide clear evidence for customers of the good value their offer represents.

They will provide help in finding the necessary finance on terms to suit the customer's situation and the cost and income streams arising from the new construction. The arrangements offered are likely to include options by which the customer owns the new facility immediately, or at some specified time in the future, or that the customer leases the new facility.

The industry commitments will be entered into by a large construction firm or, more usually, by an organization formed of several firms cooperating together in a strategic partnering arrangement. They will take the risk that they will be unable to provide what the customer needs within the agreed price. However, with established mainstream work this risk is not great: the construction organization will have firm price agreements with all its key supply chains based on well established ways of working. Also, most mainstream work is carried out quickly, which minimizes the financial risks.

The main commercial risks for firms involved in mainstream work arise when they fail to get a sufficient volume of work to cover their fixed costs and provide reasonable profits. The more capital intensive their work, the greater the risk. So their strategic financial decisions are largely concerned with balancing the benefits of investing long term in efficient production, better products and more sophisticated services against the costs of making these investments.

New-stream structures

The construction industry needs to be able to offer customers new-stream products and services. This is because some customers have individual and unusual needs that cannot be met by mainstream answers. Also established answers become obsolete when markets or technologies undergo fundamental change and so they are replaced by new-stream products and services. Some of these will develop into the next generation of mainstream answers but, initially, they have all the characteristics of new-stream products and services.

Some mainstream organizations establish new-stream divisions internally but most new-stream organizations are independent. They may have links to mainstream organizations to help them tackle difficult problems or to provide a wider training experience for their staff but people who feel at home in new-stream organizations tend to value their independence.

This is certainly the case in the UK construction industry which produces many highly creative designers and managers. For example, a high proportion of the world's most famous architects are British. At their best they produce world-class new-stream products. The management techniques that have grown up around their work are designed to support creativity by independent professionals. These characteristics form the basis of the problem-solving organizations that produce the innovative construction described in Bennett (1991).

Creative teams

Creative project teams are the essential organization units in new-stream work. New-stream projects result in relatively high-level work because the variables that have to be considered are ill-defined and the required technologies may well need to be developed. This means that new-stream teams should be formed by people who are experienced in the type of work that appears most likely to meet the project's objectives. They should be confident about their technical knowledge but be willing to try new answers, reflect on them and try another new idea in the light of the first attempt, and so on. They need to have the tenacity to continue this 'reflection in action', as Schon (1983) called it, until a good answer is found. New-stream projects will not succeed if they are entrusted to inappropriate people, no matter how experienced or skilful they may be on other kinds of project.

The kind of creativity required for new-stream construction work is not an individual activity. It comes from teamwork. It is therefore important that within a new-stream project team there is someone to play each of the roles described by Belbin (1993). The work is more likely to produce good answers if there is a well developed teamworking culture. In particular, the team members should be good at communicating their decisions and judgements in ways that other disciplines will understand. This means that architects should be able to express their ideas in terms that make sense to engineers, construction managers and specialist contractors. The same is true for all the specialist disciplines. It is difficult for teamwork to flourish sufficiently well for new answers to be found unless everyone works to ensure that communication is effective.

It is often best, for key aspects of new-stream projects, to appoint clusters of firms who have worked together successfully on earlier projects. Indeed the best results come from the same people working together on project after project. When the shared intensity of creative teamworking is repeated and repeated again, people learn to communicate at very high levels of efficiency and accuracy. They develop rich languages that allow them to discuss complex and highly abstract concepts, safe in the knowledge that they all understand the implications. A reference to a problem they solved on a previous project can convey more understanding than months of debate and argument by people working together for the first time.

The nature of new-stream work makes it unlikely that entire project teams can be kept together. So the key aspects of the project should be identified and then every effort made to allocate them to people used to working together. The key aspects could, for example, include the conceptual design, the structural elements, the external cladding, the services or the facilities management. The choice of which aspects of the project are central to its success is determined by the project's objectives and the surrounding circumstances.

Forming the project team from clusters of people used to working together has many benefits in efficiency. This is especially the case where the construction professionals have worked with the customer before. However, even when every effort is made to form project teams from clusters of people used to working together, new-stream projects nearly always give rise to some new relationships. An important reason for this is that new-stream project teams should be creative in dealing with issues and problems as they arise. So they need new ideas which may be difficult to find in well established teams. Bringing one or two new people into a team can directly provide new ideas and also serve to galvanize the established members into thinking differently about their work.

Given that new-stream work produces new designs, new ways of working and new products and services, the people who form project teams must have the confidence of their firms so that new ideas agreed by the team will be supported. This means in practice that firms must define the parameters within which their people can work. These need to include generous budget and programme limits, broadly based risk assessment procedures and very few rules about working methods. They should include stringent rules about quality and safety and have a strong ethical foundation. With these in place, it is important that individuals working in creative new-stream teams have the unconditional support of their firms, provided they stay within the defined parameters.

When people are working together for the first time, it is sensible to spend time working through a number of exercises and discussing the outcomes before they begin work on the project itself. For designers this may involve producing a series of answers to a tricky design problem and describing the steps they are taking and the issues they see as important. Repeated exercises of this kind can rapidly build a good level of understanding so that when real design problems are being tackled, the team can concentrate on finding the best answers. Similarly there are many games that help managers develop understanding and provide a language that helps them deal with real life problems. Induction courses for specialist contractors involved in the same one work stage often include exercises that ask them to role play the detailed construction planning of some aspect of their joint work. Subsequent discussions of how each of them approached the task can provide a good understanding of each other's interests and concerns which helps when they are dealing with real problems.

Exercises cannot build the almost instinctive understanding that comes from working with other people over many years. They do, however, provide a basis for understanding each other and working together effectively. There is considerable experience to show that using these kinds of training exercise increases the chance that new-stream project teams will find real innovations.

Levels of teams

Construction projects require the work of many hundreds of people. Certainly there are far too many involved to work as one single team. Best practice identifies at least four kinds of team that undertake distinct levels of work and bring distinctive skills and knowledge into new-stream project teams. *Figure 5.5* illustrates the main parts of the resulting structures.

Core team

First is the core team who determine the overall project objectives and the broad working methods to be used by the whole project team. They make the key design and management decisions and provide strategic coordination for the work of all the other teams involved. Typically core teams meet at least once a week to provide consistent overall direction for the project.

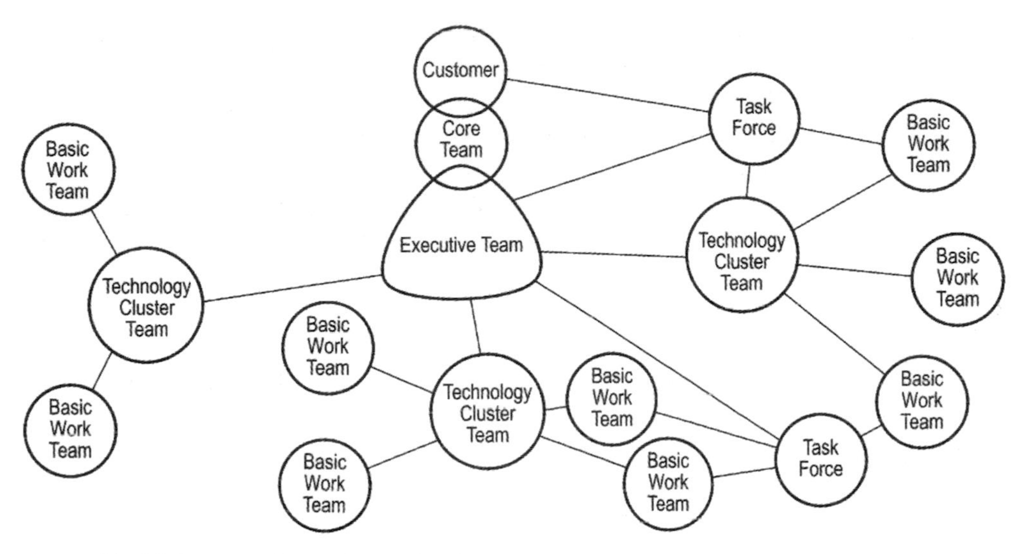

Figure 5.5: Main parts of new-stream structures

The core team brings together all the key interests so that decisions are well considered. The individual members usually include, from the customer, a gatekeeping project manager, representatives of the finance department and the users, and the facilities manager who will be responsible for running the new facility. They may also include representatives of the new facility's neighbours and the ultimate customers. The construction industry members usually include the key designers and managers; on a building project this would normally mean the concept architect, structural engineer, services engineer, construction manager and whoever has senior responsibility for the financial arrangements, either a quantity surveyor or a construction manager. Also, key specialists are usually included in the core team when their work is central to the project. This might, for example, be a groundworking specialist for a project with extensive underground work and difficult ground conditions. It might include the cladding specialist on a project where the external appearance is crucial and unusual or where there are major noise problems to be solved or some similarly demanding design problem affecting the external envelope. Information technology specialists are often brought into new-stream core teams because their skills and knowledge are increasingly central to the design of innovative products and services.

Executive team

Second is the executive team which comprises the design and management professionals who develop the project's objectives into a detailed design for a new facility and the necessary supporting services. They use the broad working methods decided by the strategic team to run the quality, time and cost control systems needed to produce the new facility and services efficiently. They provide day-to-day direction for the project by establishing a coordinating framework of designs, budgets and programmes for the specialists doing the direct work.

On a building project, the executive team is likely to include architects and engineers responsible for production drawings and specifications, managers responsible for the programme and budget, and designers and managers from the specialist contractors whose technical expertise is essential to producing practical plans for the construction work. The executive team works closely with the core team to ensure that all the key interests are taken into account and everyone involved is informed about progress and major decisions.

Technology cluster teams

The third kind of team is responsible for the technology clusters that form the systems which make up the end product. Exactly what systems and therefore which technologies will be needed can usually be determined only as the project progresses. As it becomes clear what is needed, a technology cluster team is set up to bring together key designers and project managers with managers from each of the specialists involved in direct construction work that is closely related, either by the design or the construction work.

The technology cluster teams turn the designs, budgets, programmes and other plans produced by the executive team into detailed practical plans for one major element of the construction work. A practical issue often ignored is that design interactions do not exactly mirror construction interactions. The design of engineering services in a high-rise building, for example, may interact with the foundations, structure, cladding and internal divisions. In the construction stages they may interact only with the finishing work. Both of these technology clusters should exist at the appropriate stages. They define the work of the fourth kind of team, those undertaking the direct construction work.

Basic work teams

The individual firms undertaking the basic work appoint a manager to coordinate their contribution to the project and represent their interests in the appropriate technology cluster team. These managers are involved in all the decisions about their firm's work, especially its interfaces with the work of other specialists. Thus they are closely involved in the design process and are often based on site whilst construction is underway. The nature of new-stream work means that, if they are not to delay and frustrate progress, they must have the authority to commit their firm to new ways of working.

New designs and new technologies often throw up problems that can only be dealt with on site where all the information exists. In best practice many small problems are resolved by the first-line managers doing deals with each other. When a problem impacts design, budget or programme criteria, experienced first-line managers bring their own firm's manager into the discussions. When they cannot find a good answer within the technology cluster team, they involve the executive team. In really difficult situations, the core team may become involved. This is why it is common for technology cluster teams and executive teams to be based on site for new-stream projects. They can then ensure that answers to problems agreed by the first-line managers of the basic work teams stay within the criteria established by the overall design and management systems.

Network structure

It is tempting to see this structure as a simple management hierarchy. It is more useful to think of it as a network of teams, each contributing distinct skills and knowledge essential for the project's success. Creativity is encouraged by rich patterns of interactions between committed and competent teams. The problem with a hierarchical structure is that it tries to predetermine and restrict interactions. It is not possible to work out before a new-stream project begins which interactions will trigger the best new ideas. Given competent and well motivated teams, it is best to allow them to form whatever interactions they believe are needed in order to meet the project's objectives. This is more efficient and more likely to produce successful outcomes than attempting to predetermine a fixed structure. Therefore the pattern of interactions in any real life new-stream project is likely to be different from that shown in *Figure 5.5*.

Task forces

In addition to the four centrally involved kinds of team, there is a fifth that is often used by project teams dealing with difficult problems. This is a task force set up to find an answer to a specific problem or to study a defined aspect of the project. Task forces are often an important source of entirely new ideas. Issues they might tackle include identifying ways in which the new facility could help users or neighbours. Another might be to define some aspect of the style required in a new building, perhaps to reflect the culture of the customer's organization. Another might be to study alternative foundation designs to deal with difficult ground conditions. And yet another might be to study some aspect of quality control to help the team ensure that performance requirements will be met. There are many such subjects where task forces are likely to provide good answers.

Typically, task forces bring together a few members of the project team who have a real interest in finding a good answer to a specific problem. It often helps to add one or two external experts to guide the search into new areas. Task forces work best when they have just one specific task, a relatively short time to complete the work and sufficient resources to search widely for the best possible answer. Setting up a task force can provide an effective way of moving forward when a project team is stuck on a difficult problem. It allows the rest of the team, not forming part of the task force, to make progress in other areas of the project work.

Financial arrangements

The free-ranging nature of new-stream work raises financial issues that traditional practice deals with badly. Indeed many of the construction industry's usual practices cause people to concentrate on financial matters and so make it difficult for them to work effectively, still less find innovative, new ideas. They are forced to worry that they may not be paid fairly for their work and so they spend time assembling evidence to substantiate a claim or looking for opportunities to cut their costs, even at the customer's expense. These actions are fundamentally wasteful and inefficient and many of the improvements in performance reported, for example, in Bennett and Jayes (1998), come from eliminating these non-productive activities.

The most effective financial arrangements for new-stream projects begin with the customer's business case. This establishes the value provided by the new facility and so the maximum price the customer can afford. The price will relate to defined functions and specific performance levels. The industry needs to be able to guarantee to deliver a facility that accommodates the functions and achieves the performance levels for a price that matches the customer's business case. In addition, the industry should undertake to look for savings against the agreed price that will be shared with the customer. This gives the customer an incentive to help in searching for savings.

The industry's commitments will be entered into by a large construction firm or, more usually, by an organization formed of several firms cooperating together in a strategic partnering arrangement. They will take the major risk that it could prove impossible to provide what the customer needs within the agreed price. The financial arrangements used to foster and support effective new-stream working will operate inside the construction firm or the strategic partnering arrangement and in any links they form with other construction firms.

That is the ideal arrangement but in many cases at present, customers do not have sufficient confidence in the construction firms they employ to leave these financial arrangements to them. They therefore accept the major risk themselves and set up financial arrangements which bring them fully into the overall partnering arrangement.

Whichever approach is used, as *Figure 5.6* shows, the value to the customer becomes the overall budget for the work. In the best approaches, the next step is to agree a fair profit and contribution to fixed overheads for each of the firms involved.

The calculation of a fair profit is relatively straightforward. In the early stages of any joint arrangement, it should be based on what is normally achieved by competent firms undertaking the particular kind of work in whatever is the appropriate local market. As firms work together over several projects and become more efficient because of their joint working, they should all achieve higher profits. Their greater efficiency still allows them to charge customers lower prices so that everyone wins.

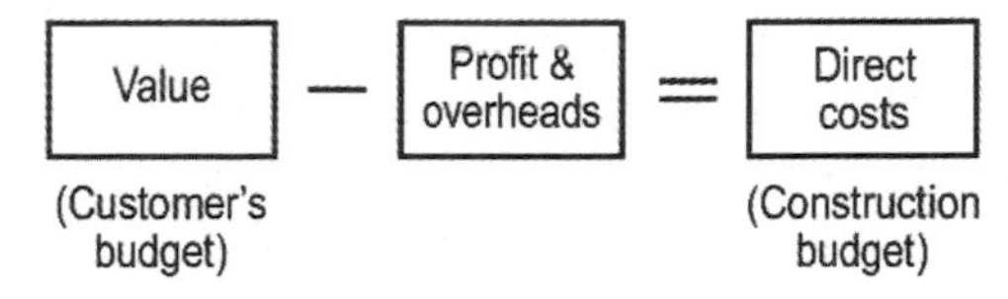

Figure 5.6: Basic financial equation

The calculation of a fair contribution to fixed overheads can be more complicated. For example, there may be costs involved in developing cooperative ways of working and these need to be taken into account in agreeing how the overall budget is to be dealt with. These costs arise principally from the need to change established ways of working. Specific costs include training in cooperative ways of working and such things as the development and use of cost control systems that support steady, continuous improvement. Also, workshops that bring teams together to build or reinforce cooperative ways of working are most effective when they are held in purpose-designed conference facilities and employ an independent facilitator. The cost of these aids to effective work are a good investment but the teams involved need to agree how they are to be paid for. It is fairly normal to include them as a charge against the overall budget but, when these training and development activities have long-term benefits, they should be seen as part of firms' fixed overheads. In these cases individual projects should be charged with only a fair proportion of the costs involved.

Once a fair profit and contribution to fixed overheads is agreed, it is guaranteed. This quite literally means that all the firms involved will get their agreed profit and contribution to fixed overheads. In addition they will be paid all the direct costs of doing their best possible work for the project. The intention of these arrangements is to remove individual money worries from the team and empower everyone to concentrate all their efforts on finding the best possible answers.

Having agreed a fair profit and contribution to fixed overheads for all the firms involved, as Figure 5.6 shows, these sums are deducted from the budget to establish the maximum cost that can be afforded. The design and cost plan are developed to meet this maximum cost by the core and executive teams. This should be done in ways that allow all those involved to understand what they are expected to do within what cost targets. Bennett and Jayes (1998) report cases where an initial cost plan is established, it is discussed in detail by the project's core and executive teams at a workshop. Everyone has the chance to ask questions until they are confident that they understand all the assumptions and decisions built into the plan. Then the cost plan and the budget on which it is based are formally accepted by everyone present. It is important that this includes the customer's key representatives to impress on them that the overall budget is fixed and no more money is available unless circumstances force their main board to make a major change to the business case. Therefore the customer's representatives know they cannot ask for additional features or add new requirements unless they are prepared to join in a search for savings elsewhere in the new facility or go to the main board and ask for more money. The formal acceptance of the cost plan and budget also serves to impress on the construction members of the team that the budget is fixed and they have to find answers that provide everything the customer needs within it.

Then as work progresses, the firms involved are paid all the direct costs of undertaking the work on the basis of open book accounting. The accounts are subject to tough audits and anyone found cheating over their accounts is removed from the team and charged with the costs involved in replacing them. In this way the audit procedures provide a very tough incentive for firms to ensure that their accounts are fair and accurate.

Cost control is used to actively search for possible savings and to identify areas where the agreed cost plan is under threat. Detailed cost reports are reviewed by the core team every week and decisions made about all cost threats and opportunities.

When, despite every best effort of the team, a cost target is exceeded, it is the responsibility of the whole team to search for savings elsewhere to bring total costs back under the overall budget. It is important that the core team includes representatives of all the customer's key interests so that savings are not made at the expense of their use or running of the new facility. Experience of using this approach on construction projects shows it to be extremely effective at motivating teams to find the best possible answers at the lowest costs. Bennett and Jayes (1998) report cases of one-off projects where the costs have been reduced by as much as 40 per cent compared with traditional approaches. Also, the timeframes have been cut by similar amounts and the quality of the end product is significantly better than normal. The people involved reported that they found the experience of being allowed to concentrate on doing their best possible work, without having to worry about their own financial position, incredibly motivating.

Evolving financial arrangements

It often takes time to build up the systems needed to adopt this best practice approach to the financial arrangements. Firms are understandably cautious about departing too far, too quickly from traditional methods which, despite their obvious faults, have enabled them to stay in business. So firms setting up new-stream organizations typically work through a number of stages before they feel sufficiently confident to adopt fully open book approaches. *Figure 5.7* shows a common evolution of the financial arrangements, providing progressively greater incentives to improve performance.

Many construction firms involved in cooperative arrangements with major customers regard the possibility of repeat business if the project goes well as sufficient incentive. Where this is reinforced by some assurance of a fixed amount of work over a period of time, firms are willing to invest in improving their own performance.

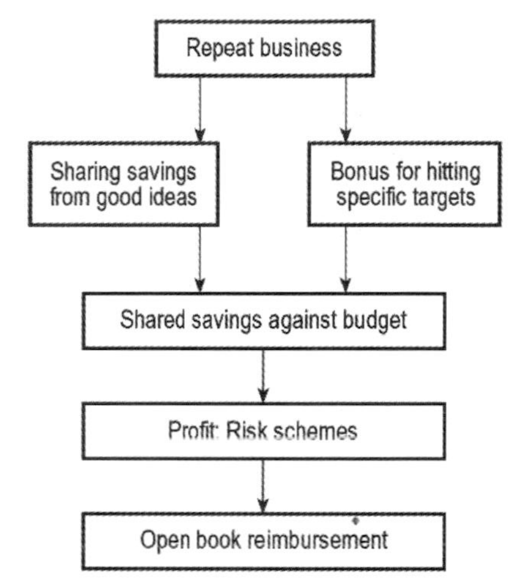

Figure 5.7: Evolution of financial arrangements

However, not taking the opportunity to improve on the industry's traditionally low profit margins is to miss further and bigger benefits for everyone involved. The essential starting point for developing new financial arrangements is to be clearly aware of the net benefits that cooperative working provides. These benefits provide scope to use monetary or non-monetary rewards over and above those available from traditional arrangements. The way these additional rewards are handled has a direct effect on the performance achieved.

A crucial requirement is that the arrangements are clearly seen to be self-financing. Therefore a simple first step that is easy to introduce, and which seems to be acceptable both to customers and construction industry firms with little or no experience of working cooperatively, is to concentrate on looking for savings and then sharing the money saved. Another approach is to set a cost or time target and then pay a bonus to the team responsible for the work if they achieve it. In one of the case studies reported in Bennett and Jayes (1998), the customer set a target cost per square metre for a new building and paid each main design consultant a lump sum of £25 000 if it was achieved. In the event, the target was missed by a small margin but the customer paid the bonus anyway. He did this to demonstrate his commitment to joint working. It may have been more effective to have not paid the bonus and so reinforced the idea that targets must be taken totally seriously on future projects. But he felt that the consultants had tried very hard to meet a tough target and deserved the reward, which he could afford from the savings they did achieve.

A more comprehensive approach, which allows project teams to work cooperatively without changing normal financial methods, is to establish the customer's overall budget for the construction work. Then time and effort is put into agreeing a fixed price for the work of each of the contractors responsible for the construction elements so that in total they fit within the budget and, if at all possible, provide savings. The prices are established at a point in the project when all costs and risks can be evaluated and must include a fair profit. Thus, each contractor's fixed price should be agreed only when he has had time to fully understand the design, including other work that may impact on his activities, to evaluate the site and understand the criteria that will be used by consultants in approving his work. The contractor is given time, often several months, to identify all the risks and agree with the rest of the executive team how these risks will be managed so that the fixed price fully allows for the agreed way of working. This approach needs an agreed method of valuing changes so they do not undermine the financial agreements; however, the intention is that there will be no changes because the customer does not intend to alter anything.

In effect, this approach achieves the benefits of cooperative working before the construction work is undertaken and then provides the security of a fixed price that is unlikely to change. The cost of these benefits is that several months are set aside for the executive and technology cluster teams to cooperate in reaching detailed agreements about every aspect of the work. Then these agreements are reflected in firm prices and the work is carried out as agreed. Any savings against the customer's original budget are shared between all the firms involved on some pre-agreed basis.

The next step in developing financial arrangements is to share savings and cost over-runs. The first stage in setting up these profit:risk sharing schemes is to establish the target price. Where the contractor has been selected by competitive tendering his tendered price is the target price. Where firms are selected on the basis of past experience or because they are part of a strategic partnering arrangement, the target price is negotiated between the customer and each of the firms involved. Ideally this negotiation is carried out early in the project so there is maximum opportunity for the project team to look for savings. In other words the target price is negotiated on the basis of the customer's business case.

The target price must be fully defined in terms of what costs it covers and what it excludes. In general, costs which are not under the control of the firms in the profit-sharing scheme should be excluded.

If the customer changes the nature of the project, the business case is altered and the target price should also be adjusted. Where the customer makes changes that are within the original business case, or problems arise for which the contractor is entitled to be paid, the target price should never-the-less remain the same. This is because within the spirit of cooperative working the project team should be committed to the success of the project from the customer's point of view and so be willing to search for savings that offset the additional costs of bad ground conditions, design problems, failure of a subcontractor, storm damage and so on.

The profits to be shared result from savings suggested by the firms involved and agreed by the core team. The actual savings are calculated under the normal contractual rules for valuing changes, taking into account any consequential additional costs. Other sums to be shared may result from an agreed lump sum incentive, which could be for completing projects early because, for example, the customer's income and profit stream begins earlier. These incentives should be agreed at the outset and should be self-financing.

An important feature of profit:risk sharing schemes is that they can minimize the impact of problems. By not requiring individual firms to carry risks arising from their joint work they do not have to include contingency allowances. Risks, by definition, may or may not occur and so rather than ask firms to include an allowance in their prices for future possibilities it is usually more effective for the whole team to have a real incentive to solve problems that do occur at the lowest possible cost.

Profit-sharing schemes also have to take into account the possibility of losses as the target cost may not be achieved. Best practice links the sharing of profits and losses. Various approaches to sharing the profits and losses are used in practice and agreeing the exact method can be an important stage in building trust within the core team.

An effective approach to agreeing the share of profits that the firms involved, including the customer, should receive is to invite each of them to state the maximum amount of profit or loss they wish to be allocated. In the early stages of developing cooperative ways of working, this may mean that the customer or the main contractor takes a large percentage of the risk and shares the remainder with the other firms involved. If the project is a success everyone gets a share of the savings made in proportion to the risk they were willing to take.

The simplest arrangement is for the maximum profit and loss to be the same figure. However, to provide greater incentives for relatively small firms who may not be able to carry a large risk, it is sometimes decided to allow a ratio of up to 2:1 in profit:loss. Any larger differential is usually regarded as commercially unsound.

Once the details of profit:risk scheme are agreed, it is sensible to discuss the whole scheme at a workshop of the core and executive teams to ensure that everyone understands how it will work. This also provides an opportunity to make sure that each firm accepts that it has been allocated a scale of rewards and risks that fairly reflects their role and responsibilities in the project. Any allocation out of line with a common-sense view of the correct balance should be discussed so that everyone understands the reasoning behind the agreed scheme. At the end of the workshop there will be an agreed list of maximum allocations of profits and losses. Actual profits and losses are then shared in proportion to the agreed list. An important implication of these approaches to providing incentives is that either the customer or the construction organization receives profits and carries losses outside of the agreed maximum allocations. Obviously this should be decided in the initial negotiations over the project. Construction organizations should be able to carry these risks but, as described earlier, some customers prefer to do so themselves because of the greater control it provides.

When a project has been completed using profit:risk sharing, the scheme should be reviewed and consideration given to how it should be developed further for use on other projects. Areas that should be considered include bringing more members of future project teams into the scheme. It can be helpful to reinforce a key objective by using lump sum incentives. A fixed sum of money paid only if the target is met can provide a powerful motivation for a team. A further development that should be considered is to base the profit or risk to be shared on actual costs rather than on the contractual calculation of the price. This can result in greater incentives for firms to search together for the most efficient ways of joint working. In this way, profit:risk sharing provides incentives to search for the most efficient answers by eliminating the need to generate claims or be defensive about working methods.

Not all firms want to take part in profit:risk sharing schemes. One way of providing such firms with an incentive to produce good ideas is for a fixed percentage of each saving to be allocated to the firm that had the good idea. The case studies suggest that a sensible figure for such fixed allocations is 20 per cent of the saving. The actual percentage should be agreed by the core team at the start of the project. The fixed allocation is paid as soon as the good idea is accepted by the core team and it is paid irrespective of the overall financial status of the project. This is because, even if the project is in a loss-making situation, it is still worthwhile to encourage firms to look for savings. A further advantage of the fixed allocation is that any firm unwilling to carry risks, and therefore not entitled to a share of the profits, will have the reward of the fixed allocation for any good ideas they provide. This may suit small design consultants, for example.

Cooperation aimed at improving performance is encouraged by this development of profit:risk schemes into the completely open book reimbursement of all direct costs plus fair profits and fixed overheads. Traditional contract clauses based on a win:lose view of financial arrangements merely create conflict which prevents firms from working efficiently.

Another common barrier to cooperation in traditional practice is professional indemnity insurance that requires firms to adopt defensive attitudes when problems arise. To prevent individual professional indemnity insurance policies getting in the way of cooperative working, project teams need to be able to accept joint responsibilities. One way of achieving this is for project indemnity insurance to be taken out by the customer for the whole of the project team. This helps creates an environment where project team members can work together for the good of the project without having to worry about their individual insurance positions.

Links to mainstream

The benefits for mainstream work of having links with new-stream organizations were discussed earlier in this chapter. However, new-stream organizations also benefit from these links.

First, they provide a source of work. This may arise when a mainstream organization concludes that a customer's particular needs require new-stream work. It may be that some aspect of a mainstream project needs a new answer and a new-stream organization is commissioned to help find it. In other cases the mainstream organization will conclude that a whole project needs to be treated as new-stream.

One advantage of these kinds of links is that larger mainstream organizations may be better able to ensure continuity in their work-load than smaller and perhaps narrowly specialized new-stream ones. A well developed link may include an understanding that the larger organization will try to help the smaller maintain a reasonable work-load so as to keep them in business.

Second, mainstream organizations can often provide the plant and equipment, well developed control systems, technical expertise, access to finance or a well founded research institute needed by a smaller new-stream organization for one specific project. This kind of link often develops into a two-way arrangement where new-stream organizations provide design talent, access to specialized computer systems and people who know how to use them, and other distinctive resources. These two-way links can form part of training schemes and generally serve to broaden the experience of staff in both organizations.

The nature of the links will vary but essentially, like all human relationships, they need open communication. A good way of ensuring this is for the two organizations to have representation on each other's main boards. This provides clear assurances for all those involved at other levels of work that the links are to be taken seriously and that communications are to be open and honest.

The main benefits of forming cooperative links between mainstream and new-stream organizations is that together they can serve customers better than either could alone. Handled well, so that customers are provided with competent construction teams whatever their specific needs, the links enhance the construction industry's reputation. Having customers who are happy and confident to commission new construction work is the best guarantee for the future of the whole construction industry.

Processes

6

Learning organizations

Processes are the coordinated actions of workers aimed at producing products and services. The new paradigm guides workers at all levels into working with, and reinforcing the natural processes of, self-organizing networks. As a result individuals, teams and whole organizations learn how to work more efficiently and produce higher levels of quality.

The actions of individual construction workers are made more efficient by training, constant familiarity and practice. Settled teams develop through trial and error, the accumulation of good ideas and all the benefits that come from shared skills and knowledge. Teams eventually become integrated units in which joint actions are virtually instinctive.

When sequences of actions undertaken by several teams are repeated, the teams develop strong links. Each team learns how the others use information, how they undertake their work, how they deal with problems and generally how cooperative they are prepared to be. This beneficial repetition results in well established processes.

Ideal processes

Ideally all construction actions contribute directly to products and services customers need, and interactions form part of well established processes that teams apply instinctively. This simplicity, however, is an ideal that will rarely, if ever, apply in every respect to any real construction work. Never-the-less it is an important model to have as a goal.

The most important factor that makes it virtually impossible to apply the ideal model is that new construction affects many people including, of course, the customer. This means construction teams face many interactions, some of which are likely to spring surprises and provide new situations for the teams to deal with. As a result construction workers face a range of situations, from the ideal model of well established processes to new types of work that involve first-time interactions between teams. A number of distinct management tools exist to support the work of teams in these various situations. In planning their work, teams must identify the actions and interactions needed and, in addition, decide which management tools to use.

Management tools

The first management tool for teams to consider is based directly on well-established processes codified into procedures. These are predetermined actions for workers to take in given situations and so, in effect, procedures define standardized processes. They are closely linked to standardized products and services, commonly referred to as standards.

Procedures and standards may be formally written and approved by firms, industry bodies or government. Or they may merely be implicit in teams' work. In whatever form they are recorded, procedures and standards tell teams how to undertake their work and how others should behave in given situations. This gives well-established teams security to concentrate on doing their own best work. Also knowing the contexts provided by other teams' procedures and standards gives them the confidence to look for better ways of working because they know what is needed for them to be effective.

Procedures and standards usually deal with a range of situations and so teams need to determine the precise constraints that apply to their particular work. Constraints are the second management tool that teams should consider. They define the levels of performance that teams have to meet. Construction constraints include official regulations about the forms of construction that are permitted, the methods that can be used, safety requirements and quality standards. Best practice makes constraints an integral part of processes but, where the work is relatively new to the teams undertaking it, they may need to use control systems to ensure that they are aware of and are working within all the constraints that apply.

Teams also have targets which they may have set for themselves or, more usually, which have been established to coordinate the work of several teams and so require joint agreement. Targets provide a measure of performance that teams aim to match or beat. The existence of targets means that teams require control systems. Together with targets, control systems are the third management tool that teams should consider. At their best, control systems rely on teams aiming at the target as an integral part of their processes. The nature and extent of the control systems needed

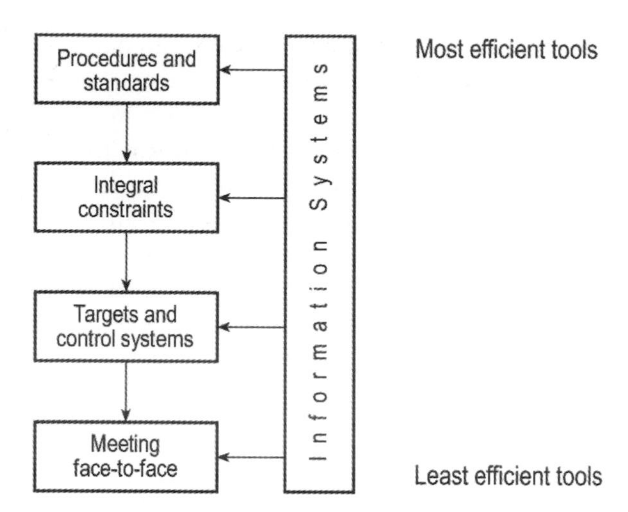

Figure 6.1: Sequence of efficiency in managment tools

depend on the team's normal performance and how close the target is to this norm. When a target is comfortably with a team's normal performance, control can be simple and infrequent. When a target is challenging, teams need detailed control systems that work in real time. A key part of these systems is that the processes and the resulting products and services are measured to provide feedback. This helps teams meet their objectives and, in the long term, provides a driving force for teams to raise the level of their normal performance.

Inevitably some situations arise that are not covered by these various management tools and teams must deal with the resulting crisis or problems by making new decisions. This may be achieved by an informal meeting, a formal meeting, a workshop or a task force. Nowadays meetings often take place at a distance by phone or by the use of some form of electronic interaction. Whatever physical form they take, these various types of direct contact are the fourth management tool that teams should consider in planning their work.

The various management tools are supported by information systems which provide the final tool for teams to consider. Good information systems provide teams with information to guide work in accordance with planned performance and to help solve problems or deal with new opportunities. The resulting framework of management tools is illustrated in *Figure 6.1*.

Planning a team's processes

Planning new work inevitably begins with meetings. Eventually targets are agreed for the required work. Only then can teams establish which constraints apply and select appropriate procedures and standards. They can select the information systems that will support their work and agree the pattern of meetings needed to supplement the other management tools they intend using.

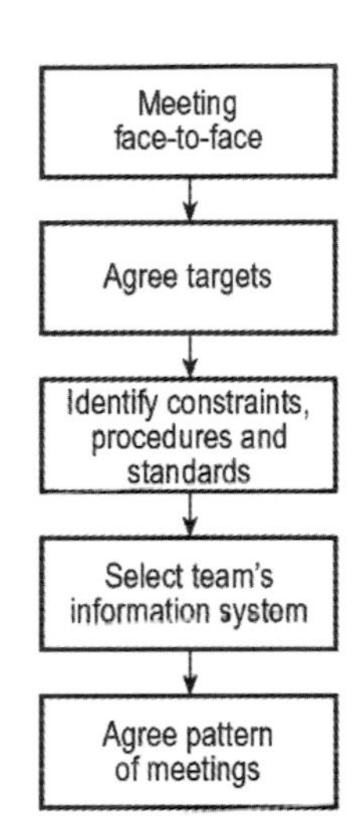

Figure 6.2: Process for planning a team's work

Figure 6.2 illustrates this planning framework which all teams can use because the ideas about processes described in this chapter relate to the work of all types of construction team: the work they do is different but all teams need to know what they are responsible for producing and how well they are doing. They need access to information when problems arise so they can be involved in finding answers. The processes that provide these various kinds of information are fundamentally the same.

The main decision to be made in planning the processes is whether the work will be mainstream or new-stream. Mainstream work relies heavily on predetermined processes. In their fully developed form they are supported by well developed information systems that enable tough targets to be set with confidence. Decisions will largely be made at formal meetings. It is likely that the work will progress in accordance with a predetermined plan and budget that are comfortably within the team's normal levels of performance. In other words, the team will work within constraints integral to their work using standards, guided by clear procedures. Achieving this level of certainty and control takes time and steady development over many projects. The effort is worthwhile because, when well developed processes have been established, they enable work to be carried out with great efficiency and certainty.

New-stream projects need flexible processes. These are directed towards fixed programmes and budgets that are interpreted flexibly in the light of actual progress as milestones are reached. This can provide a reasonable level of assurance for the customer that cost and time targets will be met, but doing so requires a tough approach to decision making. A common approach in projects using new-stream answers is to set up a project office that brings the core and executive teams together for all the significant stages. A tough cost control system is needed, often based on open book accounting and rigorous audit procedures. A clear time control system is necessary to focus the team's attention on each set of critical issues as they work through the project. Quality control has to be robust and in the hands of experienced professionals who understand the technologies being used. In best practice the core team, which includes the customer's project manager and representatives of the users, is engaged in a relentless search for the best possible answers within a fixed budget and handover date. This means that in designing the process, time should be provided as the work progresses to review the plan and make improvements.

Whatever the nature of their work, teams should jointly agree the set of actions they plan to use in producing the product and services the customer needs. Ideally the resulting processes will be familiar to the team. They should be described in process diagrams developed by the team to help everyone understand their responsibilities. The diagrams should take account of all the interactions with other teams. A basic element which can be used by teams to construct their own process diagrams is shown in *Figure 6.3*. The basic elements are linked together by interactions between teams to form a process diagram which can be as detailed as the team wishes. The links form the team's main communication, control and feedback channels. In practice the team is likely to use more links than the resulting diagram shows and it is important that they communicate with whoever they need information from. The diagram should be a guide not a straitjacket.

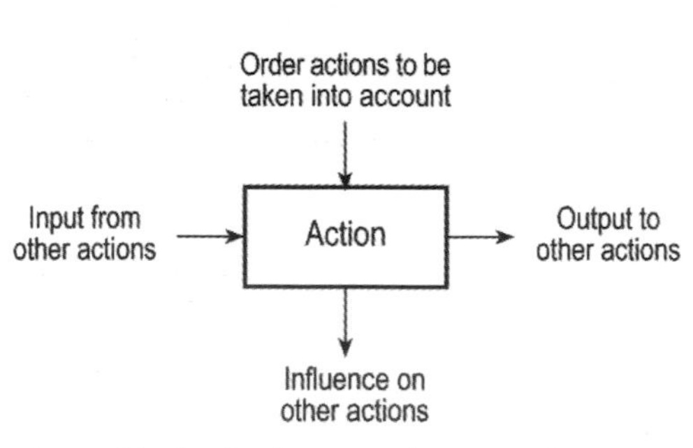

Figure 6.3: Basic elements of process diagrams

The process should be planned as far ahead and in as much detail as can be done without constraining the team in their search for the best possible answers. On mainstream projects, this should mean that the whole process can be predetermined in detail at an early stage. On new-stream projects, it may be possible in the early stages to plan only to the next milestone within a very broad overall plan. Then, as decisions are made, the plan will become more detailed.

Basic construction processes

An important stage in planning the team's processes is to identify the basic construction processes that deliver value for customers. The principle should be that the basic processes are as efficient as possible consistent with delivering high levels of quality and value for the customer.

Traditional processes include many actions that make no direct contribution to the design, manufacture or construction of the required products and services. These include work in forming contracts, defining responsibilities, obtaining competitive tenders, agreeing on responsibilities for cost, time and quality control systems, establishing liabilities and avoiding blame. These actions add no value for customers and, in that sense, should be seen as waste. A central part of modern process re-engineering techniques developed in manufacturing industries is to reduce waste by eliminating actions that add no value. This is achieved by questioning each action in the overall process:

- What is necessary for it to be carried out efficiently?
- What has to be in place before it can begin?

- Does it add value?
- How could it add more value?
- Is it essential or can it be eliminated?
- How can it be simplified?
- How can it be carried out more efficiently or quicker?
- What are the key issues, problems and risks?
- Can it be carried out in parallel with other actions?
- What determines that it is complete?

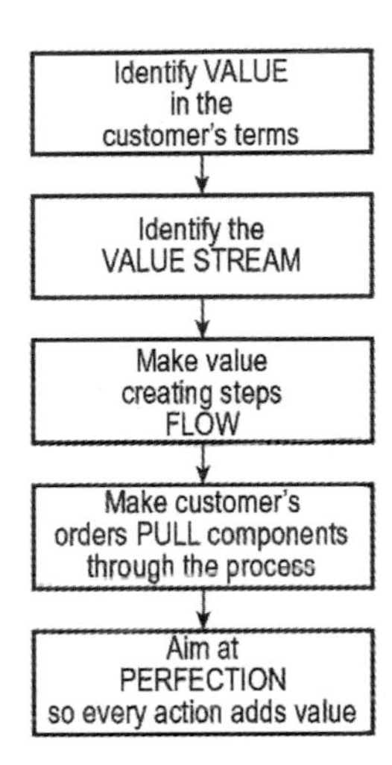

Figure 6.4: Steps in lean thinking

This systematic questioning of processes is essential for continuous improvements in performance. Quite simply, unless teams routinely question and improve the processes they use, their performance will not meet the demands of today's customers. The resulting technologies are widely called 'lean production' and, as Womack and Jones (1996) describe, are guided by lean thinking. This is applied by the steps illustrated in *Figure 6.4*.

The following descriptions of basic construction processes are intended to give teams a starting point in planning their processes. The complete set is illustrated in *Figure 6.5* which suggests the rich patterns of interactions between the basic processes. For mainstream work, the following descriptions provide checklists of the issues to be considered in ensuring that established standards provide a good answer. For new-stream work, the descriptions provide a basis for identifying the required actions. Most experienced construction professionals will already know much of this material but there may be some useful reminders for everyone, particularly about the rich patterns of interactions they need to consider.

Development

The first stages of new construction work depend on the recognition of a development need or opportunity. These arise from many sources. Much construction exists to serve basic human requirements for shelter and security. Many public-sector projects fall into this category. The needs arise from an increase in population but also from changes in the composition of a population over time. Thus, a growing

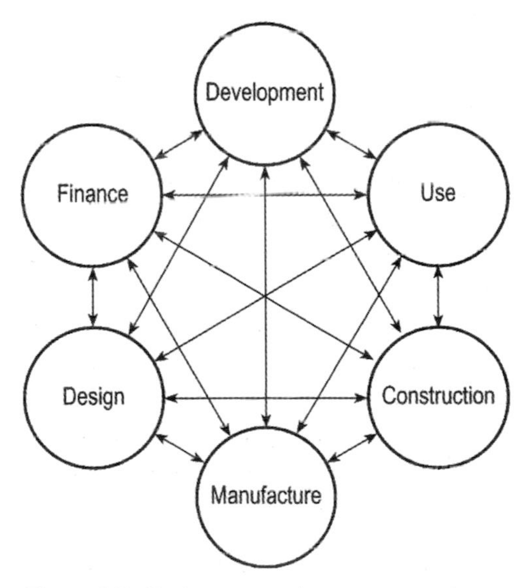

Figure 6.5: Basic construction processes

proportion of young people or of old people commonly creates a need for new construction. A different category of development needs arises from decisions to improve standards by, for example, providing better care for the old or sick, expanding educational opportunities or improving the environment.

Changes in other sectors of the economy frequently trigger the need for new construction. Thus, increases in leisure time have resulted in a wide range of new construction, especially in the field of transport. Similarly, changes in retailing fashion and habits have led to major construction projects. Deregulation in the City of London led to the so-called 'Big Bang' and to some of the largest and most important construction projects of the 1980s.

On occasions, organizations have little choice but to acquire new property if they wish to stay in existence. This may be because a lease is due to expire and it is impossible to renew it. It may be because existing property is inadequate or is beyond repair. Or it may be that competitors have invested in new technology which provides a decisive advantage and firms must respond or go out of business.

Many construction projects aim to improve the efficiency of the organizations who use their end product. A new factory usually allows the adoption of a more efficient layout of plant and equipment. New facilities can be designed to reduce running costs. A new office building may allow staff previously scattered in different buildings to be brought together. This may well result in faster and cheaper communications and even, perhaps, better decisions. A new building is often used to provide better accommodation so that staff are happier. This, in turn, can produce economies from reduced staff turnover and less absenteeism.

Many construction projects are to provide for expansion, either to meet increases in demand or to provide new services or products. Similarly, organizations moving into a new area of operation, a new business, a new location or a new way of delivering a service usually need new construction. A further factor in much construction is a desire on the part of the developer to use the new facility to enhance the image or perceived status of his organization.

A different and very important source of new construction projects is direct speculative investment. A developer sees an opportunity arising from any one of the primary sources described above and decides to undertake the development. He may build in order to sell the resulting product or he may retain it for the stream of income it will provide.

Finance

New construction usually springs from a mixture of these possible sources. The need or opportunity usually exists for some time before it is recognized and defined. However, defining a development need or opportunity is not sufficient to create a construction project. There must also be finance to fund all the inevitable costs. There are many categories of costs to be covered which may include land, rights over land, professional advice about the required permits, design, manufacture, construction, management, finance and profit.

There are many sources of finance for construction projects. A customer may use his own funds; he may qualify for a grant or soft loan of some kind from government, a quasi-government body or a charity; he may look outside for other investors through a joint venture, a share issue or bonds; or he may seek a loan. Possible sources of loans include banks, large financial institutions and major companies.

The financing of construction projects has become very sophisticated and has given rise to the expression 'financial engineering' to describe the activities of innovative financial experts. At the heart of the financial equation, which all successful developers must solve, is the fact that construction ties up large sums of money before the investment provides any return. In the early stages of a project, before there is any physical construction to attract and impress investors, finance tends to be expensive because of the risks that one or other of the multitude of things which can stop or delay a project will, in fact, occur. As projects near completion, finance becomes easier to arrange. The risks are less, most of the costs are known, in a speculative development tenants may have signed agreements to rent and, of course, the end product substantially exists. Because of these differing financial environments many projects are financed in distinct stages. For example, short-term finance is normally used to cover the costs incurred during the design, manufacture and construction of speculative developments. This usually attracts high interest rates because of the uncertainties associated with incomplete constructions. Then, as the picture becomes clearer, long-term finance is sought from investors more interested in security than in high returns. The second loan is used to pay off the first and to cover the continuing costs of running the project until the rents or other income reach a level where it becomes self-financing and, hopefully, profitable.

In all the financial decisions necessary to enable construction projects to be undertaken, the inherent risks, general interest rates, inflation rates, business confidence, tax regulations and costs will be the subject of judgement, expert advice, calculation and simple instinct. Given all the factors which could work against any construction project, it is perhaps surprising that so many are undertaken. Yet the urge to build, to leave something of oneself behind which may last for centuries, is very deep in human nature. Consequently, throughout the world many development needs and opportunities are recognized and financed and new construction projects begin.

Design

Design today is a team activity. Even simple constructions need to draw on a range of knowledge and experience greater than any one person could provide. Modern design teams are likely to include architects; a variety of engineering specialists including structural engineers, services engineers, civil engineers and process engineers; builders; specialist contractors of many kinds; managers; and of course the customers. Nevertheless, there is nearly always a single driving force behind the design of successful construction projects. This is usually a single powerful personality or, at the most, two people whose combined talents gel into a creative partnership. The single driving force is the primary designer who conceives the overall design concept. This overall vision, depending on the nature of the project, may be based on aesthetic, scientific or technological criteria. The key role of the single driving force is to maintain the original vision throughout the design processes. The driving force gives the end product a sense of integrity and coherence which cannot come from widely dispersed, democratic, team decisions. The old joke about a camel being a horse designed by a committee, although unkind to camels, does apply to construction. It may seem unfair to the many people involved with a major new building that history will attribute it to a single designer or pair of designers; but history, in attributing St Paul's Cathedral to Sir Christopher Wren or the Pompidou Centre to Renzo Piano and Richard Rogers, is recognizing an important truth about design.

The need for an individual, central, coherent vision is entirely consistent with the idea that design nowadays is necessarily a team effort. Modern constructions draw on a vast range of technologies and it is part of the skill of great designers to give coherence to the work of a vast orchestra of specialists and for the resulting product to be harmonious. In this respect the primary designer is exercising leadership within the team; as others will exercise leadership when their skills and knowledge are central to the team's work. This is how good teams work; leadership is determined by the task in hand and by the problems and issues being dealt with. The idea of a single leader, still prevalent in traditional thinking, is a mistake and contributes nothing to effective team working. The aim should be to ensure that team members respect each other's skills and knowledge but are prepared to challenge and question them while they are making decisions. Then, when a decision is agreed, the whole team concentrate on putting it into effect.

Design as a system

A systematic approach to design in architecture, described by Broadbent (1973), is based on the six main interests shown in *Figure 6.6*. Broadbent's approach is based on three interrelated systems: human, building and environment. When the human system wishes to perform actions in a particular place and the environmental system in that place is incompatible with these actions, a building system is designed to reconcile the other two. Each of these systems has two main elements which represent distinct interests. Thus, there are six main interests which must be considered and taken into account by designers.

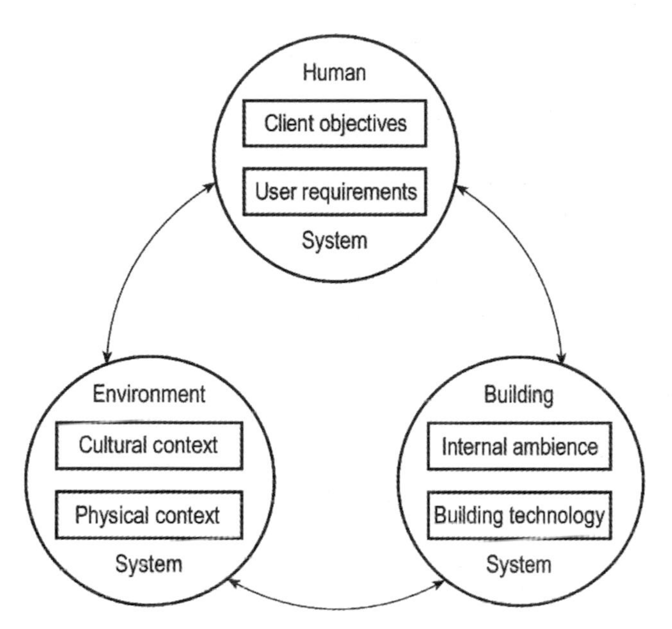

Figure 6.6: Building design system

Broadbent's analysis of the purposes of buildings is equally true of all constructed facilities, provided we add the actions of machines, plants and animals to those undertaken by people. Given this, his six main interests potentially influence all constructed facilities.

Customers' objectives

The first interest to be considered in the conventional sequence covers the development and finance issues described earlier. Broadbent brings a great concern with image to his own description of this. Indeed he implies that the prime motivation for much building is to give expression to otherwise hidden motives of the customer. He describes building as a means of exciting envy in one's competitors, establishing confidence amongst customers, advertising one's wealth, demonstrating social responsibility or showing love for one's country.

This emphasis does not deny the importance of customers' direct objectives, such as creating comfortable and stimulating spaces to house specific actions. It does, however, recognize that the environment to be modified has political, social, cultural and aesthetic dimensions as well as the more obvious physical, technological and economic ones. This produces a rich mix of objectives which need to be drawn out of customers and made explicit in an agreed statement or brief. However, there is also a need for a functional brief. This takes us into the second main interest that influences the design system.

Users' requirements

We must first recognize that the users' requirements may conflict with the customer's objectives. An extreme example is provided by prisons. However, much the same type of dichotomy arises in most public-sector building and, indeed, the distance between senior management and junior staff in almost all organizations makes a separate study of users' requirements necessary.

The aim of such a study is to produce a complete list of all the actions to be accommodated. For each we need to know the physical space required, the necessary environmental conditions, the relationships with other actions and the effects on the physical structure of the end product. Given a detailed statement for the individual actions, they can be grouped together in several ways: in terms of physical movements between actions, in terms of common environmental needs (e.g. air-conditioning, quiet and good light) or in terms of social needs. Once the clusters of actions have been identified, attention can be directed towards spatial sequences and circulation patterns. These define the processes to be accommodated. Obtaining an agreed statement of the processes is just as important as the definition of the individual actions. Just as chemicals are processed and transformed as they pass through the various elements of a processing plant, so people are processed as they move through buildings. The effects may be less dramatic but they are real and form an important part of the function of buildings. Thus it is helpful to agree a chart which illustrates the sequences in which various categories of users move into, through and out of a building.

Taken together the statements of the customer's objectives and users' requirements constitute what is usually called the brief.

Physical context

The third main interest to consider arises from analysing the proposed site for the new construction. Designers need to know what will be permitted by laws governing public interest in health, safety, convenience and amenity. They need to study the physical characteristics of the site, including its climate, geology and topography. They need to identify sources of noise, smell, visual attraction or pollution. They need to understand the social, historical, cultural and religious context of the site. They need to know if it has major political or economic significance to the local community. They should consider the opinions of neighbours as well as looking at the actual surrounding buildings and other constructions, including particularly those which provide transport or generate traffic. They need to identify existing vegetation and available services. They need to give particular consideration to views from the site and also to views of the site. The aim of all this data is to help the designers produce the outline of a desirable and permissible construction envelope. This will be represented in drawings, computer-generated representations, models or written reports on the overall size and shape of the proposed construction.

Internal ambience

The next main interest arises when the clusters of actions, spatial sequences and circulation patterns agreed in the brief are fitted into the outline construction envelope. Many of the decisions will be simple and obvious. For example, the main entrance may be determined simply by ease of access, the directors' suite may well be situated to enjoy the best views, a drawing office needs a northern aspect, production facilities may require heavy duty access for large volumes of incoming raw materials, and so on. However, as more actions are plotted into the outline construction envelope so clashes occur. Two or more actions will compete for the same physical location. At this stage there needs to be agreement on priorities in terms of comfort and stimulation for individual actions and the relationships between them. High priority actions will get the best locations. Low priority actions may have to accept poor views, noisy conditions and relatively long journeys through or around the new facility.

Having located actions within the overall construction envelope in abstract terms, it is time to consider their physical containment. This is done first in terms of the size and performance requirements for each space. We need to consider how much or how little the various actions should be separated from each other by physical walls and so on, and how far they should be open. Circulation counts as an activity and must be allocated space.

The design now consists of spaces surrounded by surfaces. The surfaces exist to contain and provide support for all the actions identified and defined in the brief. At this point in the process, the design may still be relatively abstract or it may be recognizable as a model of a real physical facility.

Cultural context

The fifth main interest arises from turning the design into a physical three-dimensional form. Broadbent makes it clear that the choice of physical form inevitably carries with it large cultural connotations, hence the title of this set of interests. He also provides detailed descriptions of four types of design process used to help in the choice of physical form.

First, there is pragmatic design. That is, working through trial and error and using knowledge about the properties of materials and the way they have been used in the past to form parts of constructed facilities. Structural forms and space divisions have traditionally been provided in mass, planar and skin constructions. Relating these traditional uses to the earlier stages of the design analysis may suggest a satisfactory shell to which will need to be added services to create the required internal environments.

Second, designers may adopt an iconic design approach. This consists of using tried and accepted forms which are repeated because they are known to work in terms of construction and the performance of the end product.

Third, designers may use analogical design. Analogy is the central mechanism of creativity. It is the process of creating a new form by using as a starting point one which already exists and playing with it. Corbusier's use of a crab shell in designing Ronchamp, Frank Lloyd Wright's use of water lilies in the design of the Johnson Wax Factory, and John Utzon's use of the sails of yachts in his design for Sydney Opera House provide famous examples of analogical design. It is perhaps more common to take an existing building or style of building as a starting point, but to use the earlier forms in an original, interesting, beautiful manner.

Fourth, designers may rely on canonic design. That is, using a given pattern, a grid of fixed proportions or some other geometric system. Such an approach is evident in the earliest architecture and remains important in schemes for the modular coordination of prefabricated components. Many of the schemes adopted in the past were based on the proportions of the human body, others are based on the geometric properties of ideal shapes, while others merely provide a regular grid or more usually a series of regular grids superimposed on each other. Designers use these to guide their choice of the sizes and shapes of all parts of the end product.

Having, by some combination of these four design processes, produced a number of possible building forms, it is wise to apply some basic checks. These can be very simple: can it be made to stand up, what will it cost, how will it behave environmentally, will it be easy to maintain and alter, will it meet the legal requirements, how long will it take to construct and does it satisfy the brief. The answers to these questions should be ranked against the agreed objectives and priorities to select the agreed conceptual design. This will be described in some combination of drawings, models and reports, although computer technology now enables designers and their customers to 'walk through' design proposals in virtual reality.

Building technology

The sixth set of main interests concern the specific technological answers for each part of the end product. This involves selecting the structural systems, the space-dividing systems, the services systems and the fitting-out systems. This is commonly called detail design and it results in working drawings and specifications.

Nowadays detail design inevitably draws on the knowledge and experience of specialist contractors. Although it is sensible as far as possible to use tried and tested methods, where new answers are needed specialist contractors provide a source of both good ideas and a deep understanding of the behaviour and performance of available technologies. Consequently designers and specialist contractors work closely together to ensure that all their interests are taken into account in the design.

Manufacture and construction

Design is not complete even when all of Broadbent's six main interests have been considered. All the details of the end product need to be designed in the light of the manufacturing and construction processes. These processes are concerned with making and assembling all the physical parts of the end product. They are divided between the site and a large number of factories which may be located anywhere in the world. The distributed nature of these processes means that managing the transport of materials and components often plays a significant part in manufacture and construction.

Standard materials and components

Teams responsible for construction projects have direct responsibility for only some of the manufacture and construction processes. Much of the relevant manufacturing industry operates independently of any one project. Indeed, a significant proportion serves other industries in addition to construction.

Many of the companies producing basic construction materials are very large compared with most of the construction consultants and contractors involved in individual projects. They are large because they produce basic materials and components that are relatively standard and so give rise to international levels of demand which provide scope for economies of scale. Indeed, many product manufacturers operate in markets where monopoly or oligopoly conditions prevail. Obvious examples include cement, common bricks, timber, steel bars and rolled sections, paint, plastic pipes, roofing felt, flat glass, concrete roofing tiles, sheet insulation, plasterboard, basic sanitary fittings and so on. For heavy materials, such as aggregates and ready-mixed concrete, local concentrations of supply also tend to result in relatively large firms.

Teams involved in individual projects are generally only marginally aware of the nature of these various material supply industries because within traditional practice they have no choice but to accept what the market provides. Or, where the supply, delivery or price performance is unacceptable, they simply avoid using the particular materials by not including them in the design of the project. The situation can be different on mega-projects or on a series of projects where an order may be sufficiently valuable to persuade a manufacturer to alter a standardized manufacturing process. However, generally, construction teams are in a relatively weak position. Never-the-less they can usually specify basic materials and expect them to be supplied in accordance with manufacturers' stated terms and conditions throughout the developed world.

The situation in developing countries can be very different. In such countries there is often no reliable supply of construction materials and managers have to buy overseas and arrange international transport, import, storage, protection and local transport. When this is necessary, these processes must be planned and taken into account in construction programmes and budgets.

Project-specific components

When project-specific components are used in either developed or developing countries, the teams involved need to plan processes which can be complex and uncertain. They need to understand the character of manufacturing so they can integrate the manufacture and construction processes.

The manufacture of bespoke products begins with a design process. Working drawings and specifications produced by design consultants provide insufficient information for manufacture. These normally provide only a general picture of the performance and appearance of the required components and specify only the materials and finishes to be used. In order to organize their manufacture, every separate part must be drawn to a large scale and all the joints and connections detailed in shop drawings. These are detailed manufacturing drawings produced by the manufacturer's production engineers and draughtsmen. It is normal for shop drawings to be checked and approved by the design consultants to ensure that the manufacturers have interpreted their design intentions correctly. It is not uncommon for designers to want to improve the manufacturer's first proposals and even their second and third. Indeed, some of the world's very best architects take this stage of design so seriously that they insist on working in manufacturers' drawing offices alongside the production engineers while the shop drawings for key components are produced. The resulting designs are the product of joint work which therefore, potentially at least, combine high aesthetic sensitivity with deep technical competence.

Shop drawings provide the basis for detailed agreements on exactly what is to be manufactured. They constitute a major activity in which, for example, a medium-sized (20 000 m^2 floor area) office building typically requires between 4000 and 6000 shop drawings. Once they are agreed by the architect, the next step in the manufacturing process is to assemble all the necessary materials and bought-in subcomponents. For even a moderately sophisticated building, these are likely to come from many different parts of the world. Gray and Flanagan (1989) report, as an example, the purchase of polished granite which was to form a major part of the cladding for a building in the City of London. It was quarried in Africa, cut and polished in Italy, and the first delivery was made, almost exactly to the day, one year after the order was placed. Only then could the prefabrication of the cladding begin. This is not an entirely unrepresentative example; indeed, the sequence of events for sophisticated bought-in subcomponents can be even more tortuous and lengthy.

Before full design approval is given, it is often necessary for a number of prototypes to be produced and tested. It may be necessary to commission original R&D work to solve unusual detail design problems. This, in turn, may require the resources of a wide variety of firms, including some from outside the construction industry. Particularly in high-tech buildings, the manufacture of bespoke components often draws on sophisticated, purpose-designed materials. As an example, Norman Foster's design for the Renault Centre in Swindon, UK, includes a detail in which the structural steel frame passes through the external envelope. Since the two elements behave very differently, the junction posed a problem. It was solved with the help of a firm whose normal business is manufacturing neoprene skirts for hovercraft. This type of interaction between construction design and ideas borrowed from other industries is becoming increasingly common. It raises new issues for construction teams who find they have to coordinate working styles and customs, some of which are quite alien to construction's normal methods of doing business. The production and agreement of the shop drawings often provides a real test of an executive team's skills.

Manufactured components

Most of the manufactured components used by the construction industry are bespoke products. The processes that produce them tend to be organized in a manner which is more concerned with flexibility and minimizing capital investment, than with speed or efficiency. This is largely a response to the extreme variety of the products ordered, combined with unpredictable patterns of demand. As a result much manufacturing remains craft based. It is slow, inefficient and expensive, but as long as all bespoke manufacturers rely on similar technologies it can be, and indeed is, profitable.

However, this world of small craft-based manufacturing is rightly being challenged. Lean production technologies have begun to demonstrate that better answers exist. In return for project teams accepting some limitations on design freedom, manufacturers can deliver standardized products quickly and reliably. There is considerable potential for manufactured components to develop much further. Indeed manufactured buildings incorporating sophisticated information technology devices to assist, entertain and support the users could well emerge as a major new global industry in the 21st century. There is a huge need for decent houses and indeed for many other constructed facilities but construction has failed to respond. New developments to meet these needs may well come from firms outside the traditional construction industry. It may be that General Motors, Toyota, IBM, Mitsubishi and similar global giants will become major players in construction in the twenty-first century leaving today's construction firms to undertake nothing more than routine site assembly processes.

Certainly traditional practice has generated a component supply industry of very variable performance. The resulting uncertainties and inefficiencies are one of the main factors causing leading customers to insist that the construction industry develops more efficient supply chains. Indeed, the Egan Report (Construction Task

Force, 1998) regards supply chains as critical to any significant improvements in performance by the UK construction industry. Modern lean production technologies need to be at the heart of the industry's supply chains if these improvements are to be achieved.

Planning site processes

The processes on site provide the direct focus for the work of teams responsible for individual projects. Best practice is for their planning to begin early with an evaluation of the site from a construction viewpoint by construction managers with up-to-date knowledge and experience of the local construction industry. Having determined in general terms the kinds of construction that could take place on the site, the construction managers agree the overall method of construction in agreement with the rest of the core and executive teams.

As more detailed designs, construction plans, programmes and budgets are produced, the construction work packages and the interactions between them can be identified. These establish the overall sequence of construction processes, and the detailed flow of materials and components around the site as they are unloaded and fixed into position.

Planned in this way, site processes should be relatively straightforward and efficient. It is the traditional approach, in which construction management expertise is involved far too late, that causes site work to be planned hurriedly on the basis of design decisions that take no account of the construction implications. The approach is made even worse by arbitrary changes to the design introduced after work on site has begun. As a result construction strategies have traditionally been designed to be flexible, not efficient or fast. This is no longer acceptable and, by understanding the rich connections between the work of everyone involved and bringing them into the team early, teams are beginning to reduce construction times and costs.

Specialist contractors' work

Whatever form it takes the construction strategy provides one of the two key controls, conceptual design is the other, for specialist contractors' work. Their work must fit within the design and construction framework so formed. Specialist contractors divide their own work into major stages and produce their own strategic method statement, programme and budget. These cover the production of shop drawings, agreement of the design, production engineering, procurement of materials and components, manufacturing, delivery, storage, distribution around the site and final construction. They plan the use of major items of plant and equipment on site, essential temporary works, on-site and off-site storage, safety, welfare, major labour relations issues and the removal of rubbish and waste materials.

It is normal for specialist contractors to be involved in coordinating their work with that of other specialist contractors, in cooperation with the designers and construction managers in the executive team. This helps ensure that the designers and construction managers are happy with, and can agree, the specialist contractors' design and construction proposals. They need to ensure that the proposals are in accordance with the overall plans or, where problems arise, answers are fully considered, discussed and agreed with all the teams who may be affected by new decisions. Time pressures on many projects mean that the set of coordinated method statements, programmes and budgets will still be under development after construction on site has started. It is only through all the teams involved working in cooperation that controlled, efficient progress can be maintained.

Managing construction on site

Construction is essentially a sequential process of assembling components which provide the support for subsequent components. Support at intermediate stages is provided by temporary works, plant, equipment and the muscle power of workers. New facilities have always been constructed in this way and, where projects use well established technologies and craft skills, regular patterns of working have emerged. In these circumstances construction managers and specialist contractors know the sequences of work, the types of plant and equipment needed at each stage, and the nature of the temporary works needed to provide support for workers and their work. Work progresses relatively slowly but there are few surprises and there is relatively little need for formal management.

However, where projects use new technologies, new materials and new skills, construction throws up many problems for managers. The difficulties arise from the absence of directly relevant experience. Partly because of this, one of the important implications of innovation is that design has to be very detailed in order to provide precise information for the construction of every component. It is not possible to rely on craft knowledge to make good any deficiencies in the design information. The construction workers may well not fully understand the detail design philosophy and therefore modifications on site can all too easily cause failures. The result is a multitude of drawings which must not only be checked and agreed by designers, construction managers and specialist contractors, but must also be understood by the workers on site so that construction follows the designer's intentions exactly.

Many modern projects combine traditional work with new technologies. This complicates the design and construction tasks even further. There are not only the very different processes required by the two forms of construction, but the work itself is also very different. For example, prefabricated components are normally produced to tolerances of the order of ±1 mm, while traditional work normally has

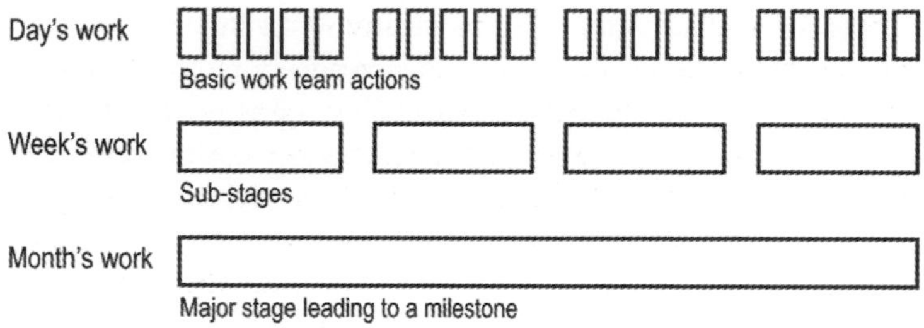

Day's work
Basic work team actions
Week's work
Sub-stages
Month's work
Major stage leading to a milestone

Figure 6.7: Natural construction planning framework

tolerances of at least ±10 mm. Forming connections between the two is as difficult as creating an efficient organization which combines the two distinct philosophies of site-based craftwork and factory-based prefabrication.

Experienced construction managers work with the normal time framework to divide construction processes into hierarchies of self-contained and self-sufficient stages. *Figure 6.7* illustrates the approach which begins by identifying the major stages which for a building project may be site organization, substructure, structure, external envelope, internal subdivisions, concealed fittings and services, finishes, exposed fittings and services and decoration, and commissioning. Each is the responsibility of a technology cluster team who divide their processes into substages which are the responsibility of a direct work team. On very large projects there may need to be another level of sub-substages. Then finally the work is divided into one-day units for each direct work team.

Best practice aims to complete the planned work for each day within that day. When this proves impossible, effort is concentrated on directing the project back onto planned progress by the end of the current substage or, for larger projects, the current sub-substage. This is normally one week's work. When this cannot be achieved, a crisis situation is signalled and every effort is concentrated on achieving the next milestone, normally relating to a month's work Thus there is a clear systematic focus on putting the overall plan into effect and not allowing time overruns to accumulate.

Traditionally it has been unusual for this degree of control to be achieved on new-stream projects. Many things can go wrong and extraordinary efforts were required for projects to be finished on time. As a result failure came to be widely accepted and regarded as normal. Standard forms of contract include many excuses for failure and in the eyes of many practitioners this legitimizes late completions, cost overruns and defects in the finished products. These attitudes have no future in a modern construction industry; they should never have been acceptable but now major customers simply refuse to accept the consequences of sloppy work.

Leading practice goes further and not only meets completion dates reliably but also works to very short timescales. This is achieved by using well established designs supported by tried and tested construction methods. As a result construction times are being reduced, as *Table 1.2* shows, by up to 80 per cent.

Even fast construction must include time to commission all the systems that form the end product. The environmental control services are the most obvious example where the proper performance of the new facility depends on all the parts working together as an integrated system. This means that all the parts need to be in place before the system can be finally tested and tuned to ensure that it works as planned. Commissioning can involve weeks of painstaking and careful work to ensure that everything is fully working. This time must be planned into the construction processes from the outset, otherwise it is inevitable that facilities will be handed over late or incomplete.

Facilities management

Construction projects have traditionally ended with the completion of the construction processes. However, some continue into a further stage which is dominated by decisions about facilities management. This means establishing or helping to establish a facilities management system to run the end product; the aims of which are to help the owner use and operate his new facility to provide environments in which people can live and work happily and efficiently.

Facilities management requires staff who are trained in the proper operation of their facilities. They need to be supported by systems which monitor the performance of the facilities and the behaviour of users. The systems need to identify significant changes and trends so that problems and opportunities can be identified. In many cases these systems are built into the fabric of the facility. In processing plants, for example, sensing devices distributed throughout the facility provide managers with up-to-the-minute information about the behaviour of every key part. The facilities managers can sit in a computerized control room and monitor the behaviour of the entire plant. As operating problems arise, the managers adjust the plant to keep it running efficiently at the required rate. Increasingly, however, these routine adjustments are built into computerized expert systems, which take the place of human control. Human intervention is then restricted to strategic changes; that is, to finding ways of improving the performance of the whole plant, for example by spotting bottlenecks and altering the plant to remove them, planning and implementing major maintenance processes and dealing with failures of components or systems.

Many large modern buildings have similarly sophisticated monitoring and control systems. Armed with the comprehensive information they provide, facilities managers can make a difference to the performance of the organizations who occupy their buildings. This is especially true where the work of the organization using the building is subject to frequent and unpredictable changes. In these circumstances, facilities managers have a major role to play in helping to maintain users' motivation by using the inherent flexibility and adaptability of buildings to create environments which support new patterns of work. This is why construction projects should run seamlessly into the facilities management stage. Doing so brings construction professionals into customers' organizations where they can develop deep understandings of how new construction can support successful enterprises. This understanding is becoming increasingly common in best practice as customers are forced by global competition to search for lower costs in every aspect of their work. Leading construction firms are deeply involved in this relentless drive for greater efficiency and the resulting close links with customers are a key feature of the practice that results from adopting the new paradigm.

Procedures and standards

Once a team has identified the actions and interactions that make up its processes, they can agree the procedures and standards that apply to their work. Ideally they will be able to use procedures and standards built up through experience over time that make management an integral part of their normal work and so provide the basis for great efficiency.

These effects were observed by a group of UK construction managers on a study visit to Japan. DTI Overseas Science and Technology Expert Mission to Japan (1995) describes how a group of very experienced mangers were surprised by the way improvements were seen in Japanese construction as technological rather than managerial in nature. They found that management is simply not an issue in Japan. Comprehensive procedures and standards ensure that all the basic construction processes are efficient. More than this, the procedures and standards give construction teams time and resources to reflect on and assess feedback in improving their processes still further.

This is one of the most fundamental ideas that is derived from the new paradigm. Enabling teams to work with the self-organizing capacities of feedback-driven networks means they can afford to look for ways of adding value for themselves and their customers. To achieve the same efficiency, UK construction teams first need to establish comprehensive procedures and standards. This necessarily begins in a mundane manner by considering the procedures and standards that already exist.

Public procedures and standards

Public procedures and standards are important in defining the objectives and restrictions that society expects the construction industry to take into account. Public procedures and standards provide agreed answers to commonly occurring problems. They are developed in one of two main ways: either one answer is clearly identified as the best and is widely adopted so that it becomes a *de facto* procedure or standard. Alternatively, a committee of experts is set up to decide on the best answer and this is published as a formal procedure or standard.

Historically, informal methods of creating procedures and standards were the most important. As buildings were built and roads and services constructed, so workers tried various methods of using construction materials. The best methods were repeated, passed on from craftsman to apprentice, became the standard approach and were used widely. The procedures and standards were gradually recorded so that new entrants to the industry could supplement their practical training by referring to textbooks. These became more important with the growth of the construction professions. Architects and engineers needed some knowledge of a range of construction details without going through the practical training which provided the basis of craftsmen's knowledge. So, in the nineteenth century detailed building construction textbooks became widely available. Architects and engineers used the textbook details in designs for new constructions; gradually they began to change the details on the basis of their experience and, even more gradually, on the basis of formal research.

Another source of informal procedures and standards which has become increasingly important over the past century, is provided by trade literature issued by manufacturers of components and materials. It is obviously in their own best interests that their products should be used successfully. They therefore publish descriptions of how to use their components and materials in various situations and circumstances. Where a monopoly or near monopoly situation exists, the literature produced by the one dominant firm becomes the effective procedure or standard. Where a sector of the market is served by many firms, it is common for them to form a trade association to produce agreed descriptions of good practice which are used widely as *de facto* procedures and standards. These are increasingly contained in computerized information systems that make it easy for customers and their advisors to select, buy and use the products. Construction teams need to be aware of these systems and make sure they are being used appropriately in their processes.

Formal procedures and standards

It is not uncommon for informal procedures and standards to be adopted as formal national ones. In general this comes about through the operation of self-interest. A simple and obvious example is provided by a committee of experts assembled to produce a national standard specification for a material manufactured by a monopoly firm. That firm will own much of the relevant practical expertise. Therefore the firm's approach is the only practical basis for the resulting national standard. There are a number of obvious temptations for firms in such a position. They may seek to embody wide tolerances and low quality in the formal specifications in order to avoid creating problems for themselves in meeting national standards. Another firm operating in a different market may seek to incorporate high standards or distinctive, arbitrary features of their own products in order to make life difficult for potential competitors. In practice it is impossible to produce procedures and standards which are free from any form of bias.

A distinctive feature of UK construction practice is the existence of major negotiated procedures. These include various standard forms of contract, standard methods of measurement and standard forms of coordinated project information. These are produced by committees made up of representatives of the various industry bodies that sponsor the procedure. The representatives are there to protect the interests of their members. As Hillebrandt (1984) observed, the members of these committees act without pay and are usually successful and therefore busy in their own fields of operation. As a result, the meetings of the committees are held at long intervals and the process of producing a new procedure tends to be lengthy and cumbersome. The slow pace of committee work is compounded by the insistence of constituent bodies on individually examining, and usually altering, every new procedure and every revision of an existing one before giving their formal approval.

As a consequence of the way they are produced, formal procedures and standards tend to describe a conservative view of current good practice. They do not represent the best; instead they provide a middle-of-the-road, majority view. This, of course, is their strength. It means that they are safe to use. A practitioner using a formal procedure and standard to solve a problem of the type for which it was produced is unlikely to be regarded as acting negligently. Construction teams should consider how far these strengths and weaknesses match their objectives when deciding to use public procedures and standards.

Organization procedures and standards

Organizations develop internal procedures and standards and these have to be taken into account by teams that form part of the organization. Some apply to all the teams that make up the organization, others relate to just some teams, whilst yet others grow up within specific teams and apply only to their work.

Organization procedures and standards may provide a standard approach to design. This may simply be the adoption of a distinctive style which marks out their products from those of their competitors. This approach is very evident in the work of many of the world's greatest architects whose work is immediately recognizable because of distinctive ways of using particular details or forms. They develop personal vocabularies of materials and styles which are refined in successive new designs but which essentially remain standard. Much of the information which describes these design standards remains in the head of the lead designer and the teams of specialists with whom he or she works.

Design procedures usually deal with more mundane, but never-the-less important, matters including how design information should be prepared, presented and checked. Modern computer-based design systems deal with many of these issues but small incompatibilities between systems used by different firms making up one team can still cause unexpected problems. Particularly where firms are working together for the first time, it is wise to check how each of their computer systems handles specific information. Different assumptions built into linked systems can cause expensive problems.

More developed forms of design procedures and standards provide comprehensive sets of design details. It is increasingly common for these to be tailored to specific building types or even to the requirements of individual customers. The technical description of each detail is often supported by information about its use, problems that have arisen in particular situations and the solutions adopted. The most advanced systems provide a mass of supporting information concerning availability, price, construction implications and maintenance information. This means that a designer selecting a particular detail automatically generates a mass of information that is directly relevant to other teams involved in the work.

Organizations also develop procedures for their production processes. So, faced with a particular manufacturing or construction task, a team will adopt an established approach. This may or may not be explicit. The standard answer may be embodied in detailed procedures. Alternatively, it may simply be used and reused as workers base each new decision on what was done last time. In this way standard repertoires of joints, junctions and fixings gradually evolve. Particular ways of forming shapes, creating distinctive colours, achieving specific performances and handling the multitude of day-to-day problems involved in making things are repeated as long as they are successful. Over time they become formal or informal organization procedures.

In much the same way, organizations develop procedures for the behaviour of their staff. In addition to the obvious need for procedures describing direct work processes, organizations tell workers how to behave towards each other; how to behave towards customers, suppliers and subcontractors; how to act in a crisis and what to do if an accident occurs; and commonly provide rules for every situation in which the efficiency or survival of the organization may be affected. Organizations need procedures for dealing with matters of discipline, for specifying criteria used in testing their products, for keeping records (particularly financial accounts), for collecting and handling information, for allocating responsibilities and for many other matters.

Use of procedures and standards

Construction teams should, as far as possible, be formed from firms that use appropriate organization procedures and standards and relevant public procedures and standards. This is particularly the case where organization procedures and standards have been developed jointly by the members of the team and embedded in information systems shared by the firms through some form of network. This often happens where firms specialize in a particular type of project in a given geographical region. The firms who build skyscrapers in Chicago are one obvious and well known example. They have developed simple, direct procedures and standards that enable them to construct high-rise buildings at a speed that makes it look as if they are growing organically from day to day.

Procedures and standards provide many benefits. They reduce the need for new information to be produced and communicated. There is less need for new design work. It is usually easy to obtain official approvals quickly. The materials and components required can be ordered early. Programmes can be based on direct experience of producing a similar product. Manufacturers can reuse their shop drawings and may even have a dedicated production line for the components. Construction on site begins some way down the learning curve and so is immediately more productive than the construction of a new, non-standard design. Work on site will experience few problems arising from design details that do not work or components that do not fit properly. All the inevitable teething problems which come with a new design will have been identified and eliminated on previous projects. Waste from mistakes and the need to redo work are eliminated or at least greatly reduced. There is every incentive to plan for fast, accurate work and to expect the results to be reliable and efficient.

- Planning regulations
- Technical regulations
- Nuisance regulations
- Working conditions
- Designs
- Specifications
- Contracts
- Employment conditions
- Discipline
- Project information
- Construction methods
- Health and safety
- Accidents
- Quality records
- Costs
- Financial accounts
- Payments
- Programmes
- Performance measures
- Communications
- Feedback channels

Figure 6.8: Basic checklist of issues that can be dealt with by procedures and standards

These benefits mean teams should use procedures and standards wherever they fit their work. Traditionally, UK construction teams, especially on building projects, have preferred to look for new answers even when perfectly good procedures and standards exist. *Figure 6.8* provides a basic checklist of issues that can be covered by procedures and standards. All too often the potential benefits of using well established answers have been ignored by teams determined to produce something different. This is one of the prime causes of the industry's weaknesses. There is therefore little risk that the industry will fall into the opposite trap of using procedures and standards where a new answer could provide substantial benefits. This suggests that the principle to be applied is to use an established answer embodied in procedures and standards unless a new answer provides measurable, significant benefits. The measurements should be objective and independent. The benefits should be in terms of better value to the customer or of a specific benefit to the construction organization.

Integral constraints

Construction teams are required to work within certain constraints. Ideally these are an integral part of procedures and standards which teams are well practised at using. The Japanese construction industry's approach to quality control provides a good illustration of such integral constraints. In this case the constraint is to achieve zero defects when a new facility is handed over to the customer. The following description provides a useful model for construction teams in the UK to aim towards. When their particular circumstances make it impossible to apply some aspects of the Japanese model, those aspects should be given explicit attention.

Quality control is an integral part of the normal way of working of Japan's major design build contractors. Their approach uses highly developed processes that are the result of many years of steady, continuous improvement. The processes are applied with great consistency by all the teams involved on individual projects but include procedures aimed at continuous, long-term improvements to all aspects of the firms' performance.

The efficiency of the processes depends on long-term relationships between customers, design build contractors, suppliers, specialist subcontractors and sub-subcontractors. These relationships operate within tightly knit families of firms that have existed for decades and rely on cooperation and trust to form highly competent construction organizations.

The design build contractors work closely with customers to identify opportunities where new construction will help the customers' businesses. This allows them to be confident in designing to the quality standards that the particular customer needs and expects. They also work closely with, and take responsibility for, the well-being of subcontractors. This allows them to set tough quality standards for construction work and invest in helping the subcontractors achieve them. Prices are set which ensure that subcontractors earn a fair profit and can afford to concentrate all their efforts on doing the work on time and to the required quality standards. The fair prices also help ensure that subcontractors are willing to try new technologies and actively look for ways of improving quality, safety and productivity. The design build contractors help them use these same cooperative methods in working with their sub-subcontractors.

The firms have consistent processes, based on tried and tested methods of working, which are used on all the firm's projects. Staff are encouraged to search for ways of improving these established methods by meeting regularly in quality circles. In addition, new ideas and potential major improvements are researched and developed in large and very competent in-house research institutes. When a new and better method is identified, either as a result of ideas developed within a quality circle or as a result of R&D, it is discussed widely with all those likely to be affected. Only when there is a wide understanding of the change is it introduced. Then it becomes the firm's standard approach until a further new and fully considered better answer is found. By providing for innovation in this way, Japanese design build contractors combine steady, reliable efficiency and quality with continuous improvement.

A key part of this highly developed approach is that it is normal for design build contractors to appoint a very experienced project manager to work with the customer from the outset of a new project. He or she guides the project through well developed processes in which quality is an integral part.

These processes begin with conceptual design which, in Japan, is nearly always produced by architects employed by the major design build contractors. There are independent design consultants and some customers employ them to produce a conceptual design for a new project; however, the more usual arrangement is for the customer's regular design build contractor to take responsibility for the complete project from the outset. Perhaps as a consequence, the aesthetic aspects of conceptual designs are regarded as a weakness of the Japanese approach by people used to Western architecture. Japanese buildings are well built but they look dull, reflecting the solid middle-class values that permeate Japanese society. People in Japan do not like to stand out and their buildings reflect this modesty.

The limitations of traditional conceptual design in Japan are now recognized and a growing number of new buildings are being designed by some of the best European designers, including Richard Rogers, Norman Foster, Renzo Piano and Arup Associates. This exposes Japanese contractors to greater variety and a different level of design quality. They deal with the resulting new situations in the design stages to ensure that, before work begins on site, the required work will fit into their controlled and productive construction methods and that quality will not be compromised by new or unexpected situations.

The groundwork for Japanese quality and productivity is established during closely integrated detail design and construction planning processes. These are applied, whatever the origins of the conceptual design, by specialist departments at head office under the detailed supervision of the project manager. The output of these processes is a very detailed plan of the construction actions which makes extensive use of standards in every aspect of the work. The plan is described in drawings showing every detail of the design, detailed method statements, a detailed programme and a budget that matches the cost originally specified by the customer.

The plan uses standardized design details and specifications based on materials and methods familiar to subcontractors. A key part of this is a comprehensive national standard specification for construction materials and methods that is used throughout the Japanese construction industry. It is revised only when a properly researched and clearly better answer is found. This standard helps ensure that every detail of the construction method for individual projects is established and planned before manufacturing and construction begin.

Construction planning uses detailed networks, bar charts and construction method drawings to provide strong visual images that help planners think clearly about every aspect of the construction process. Planning depends for its success on design build firms knowing their subcontractors through having worked with them for decades. They therefore select design details, choose methods of construction and determine the sequence and timing of every activity with considerable confidence that quality standards will be achieved. The confidence of knowing that their plans will be taken totally seriously encourages detail designers and construction planners to consider every aspect of the work. Throughout these activities the project manager seeks to eliminate any action which does not directly add value to the buildings or services that the customer will receive. In particular, any design details that would make it difficult for workers to meet the required quality standards are changed before they become a problem on site.

The project manager has every opportunity and incentive to prepare thoroughly before work begins on site. He or she is totally responsible for the project and has the support of a large, well educated and experienced team. Their task is to help the subcontractors put the agreed plan into effect. The extreme detail of pre-construction planning enables the project management team to arrive on site knowing exactly what they must do to complete the building. As far as is humanly possible, nothing is left to chance. The overall effect is that efficiency and quality, based on using standards and well practised skills, are designed into Japanese buildings during the integrated detail design and construction planning processes.

The major design build contractors depend on subcontractors to undertake the basic work. The subcontractors directly employ key workers and take responsibility for training them to meet the rigorous quality standards. However, they create flexibility for themselves by using tiers of sub-subcontractors to undertake much of the direct construction work. Relationships are relatively informal and uncertain at the lower levels, bankruptcies and unemployment are not uncommon and small firms at the base of the structure bear the brunt of technological and market changes. This helps insulate the design build contractors and their first tier subcontractors from change and makes it easier to ensure continuity so they can afford to invest in ever-more efficient and reliable ways of working.

The design build contractors make sure that their first tier subcontractors have a steady workload. They use contracts based on terms which are fair to both parties. Consequently conflicts do not arise about payments or claims for extra money. Very little paperwork is generated. Subcontractors can concentrate on doing the required work to the best of their ability without having to worry about money or workload issues. As a result, everyone on site takes quality control absolutely seriously.

Quality control

In the trilogy of quality, time and cost, Japanese customers put quality first. They expect new things to be perfect and to work properly. These tough standards are applied to buildings. Quality in Japanese building begins with the use of tried and tested methods. Project managers know the performance of the details, components and systems they use and so appropriate quality is designed and planned into Japanese buildings. This does not mean the highest quality everywhere; it means that a knowledgeable, experienced choice is made of the right quality for each part of the end product. When a new solution is essential, the contractor's research institute will be called in to work with the design team to test and develop a new answer. However, generally, detail design uses answers already familiar to the subcontractors who will manufacture and construct them.

On site, quality is measured continuously by means of detailed, rigorous and carefully planned tests specified in manuals produced by the design build contractors. The tests are carried out consistently and, to ensure that standards never slip, each subcontractor is required to maintain a photographic record of tests and their results. This provides a simple and comprehensive record of the application of quality control systems. As a result, Japanese buildings generally work exactly as designed when they are handed over to the customer. Should any defects occur the contractor puts them right straightaway and absorbs the costs. This total responsibility derives not from provisions in the contract but from a sense of pride in the work, combined with a mature commercial awareness of the importance of happy customers.

The other major reason for giving a high priority to quality is simply that the Japanese have recognized that getting work right first time is essential for high productivity and fast construction. Re-working, altering work already done, making good defects or simply having to return to a workplace to complete work left unfinished costs money, wastes time, destroys motivation and eats into profits.

The result is that zero defects is a reality. Japanese customers are amazed by the concept of practical completion included in many Western construction contracts. They have learnt to expect their new facilities to be handed over fully complete and see the idea of practical completion simply as an excuse for poor quality work. The Japanese experience shows that reliable quality control can be achieved in construction. The key is to regard the defined quality standard as a constraint that must be met, not as a target to be aimed for.

Safety
Quality
Efficiency
Programme
Budget
Open communication
Trust

Figure 6.9: Basic checklist of issues that can be treated as constraints

Figure 6.9 provides a basic checklist of constraints that should be an integral part of the work of efficient teams.

Targets and control systems

Construction teams have targets in addition to constraints. These need to be supported by control systems. The general pattern of control systems in construction work is illustrated in *Figure 6.10*. The aim of control systems is to provide teams with an early warning that work is deviating from the target, so they know where to concentrate their efforts. Control systems need clear, measurable targets, accurate and timely feedback which is compared with the targets, and action to alter the situation when a deviation is signalled. Each of these steps is difficult, which is an important practical reason for managers to concentrate their efforts on a small number of absolutely key targets.

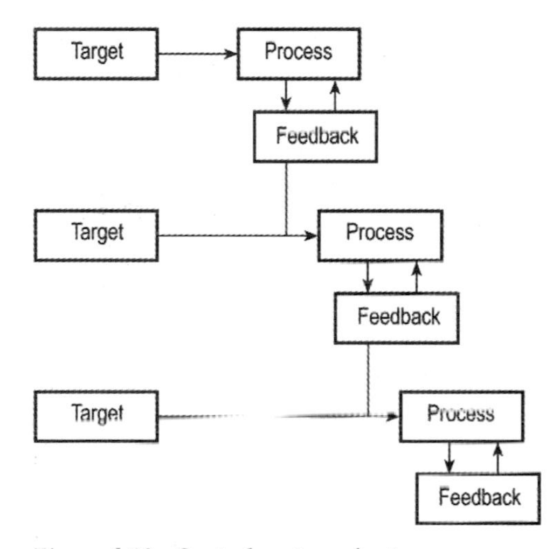

Figure 6.10: Control systems for teams

Peters (1987) emphasizes the importance of identifying a few simple, measurable targets which are of central importance to the success of the team. The measures should be active, subject to constant discussion by workers, rather than being guided by infrequent formal reviews by managers outside the team. They should be visible, shared with everyone and supported by training, so that everyone understands how they are produced and what they mean for the success of the project.

As Chapter 2 describes, benchmarks can provide reliable information about the best performance being achieved by others. There is considerable evidence that objective benchmarks of what other teams in similar situations are achieving encourages teams to set very tough targets for themselves. More importantly, they will be highly motivated to succeed, as long as it is clear that they have the authority to search for new ways of working aimed at meeting the targets.

Teams should agree how their performance is to be measured and make sure they agree the definitions of measurements as a key part of designing their processes. Where no well established targets already exist, it is sensible for teams to develop rough, unconventional, paper-and-pencil measures of their own performance. These often provide robust starting points in establishing the measures to be used. It is important that the measurements are easily understood and so can be discussed within teams, between teams and with customers.

In many cases teams will decide to use established benchmarks that relate consistently to objectives set by the strategic team. This allows everyone involved to make meaningful comparisons of the performance of individual projects. Common examples include the cost of a given unit of completed work, the time taken to complete a standard unit of construction, the number of revised design drawings issued, number of injuries per 1000 man-days worked on site, quality of welfare and safety provisions, deviations from budgets and programmes, number of change orders issued, speed of getting back on programme after a delay, proportion of rejected materials or components delivered to site, number of defects on handover, and complaints from customers or neighbours.

Broader examples include the number of real innovations introduced, number of awards given for meeting targets, hours devoted to training, time spent by teams each day on key objectives, and number of people at meetings who are directly involved in the subjects being discussed.

Another example is consultants benchmarking the time taken to respond to requests for information from other team members. Specialist contractors have used benchmarks of the time taken and the costs involved in dealing with defects. Customers have used the running costs of their buildings as benchmarks to identify significant improvements in energy use and cleaning and maintenance costs.

In deciding what to measure, teams should initially focus on those areas that will have the most direct positive impact on their performance or on their relationships with customers. Often the temptation is to try to benchmark too many things. Usually it is best to begin by measuring only the top two or three priorities. Each benchmark should aim to improve the team's performance in ways that contribute to their profit levels or to the customer's satisfaction levels.

It is important to set targets that teams have every chance of achieving and then to provide all the support necessary to ensure that only very occasionally will any team fail to meet them. Given all of this, once a team has agreed a target and been given the resources needed to meet it, there are very few excuses for failure. This means that teams must act to solve problems. When a key target is in danger of being missed, this must be treated as a crisis and clear, effective action taken quickly to get the work back on its planned course. A control system which is not used to provide this steady, systematic control is simply a waste of resources; once it is discredited by being ignored, nobody will bother to provide accurate or up-to-date feedback. Resources are wasted in producing useless measurements and, worse still, teams have no systematic warning of emerging problems. On the other hand, if a system is taken seriously, it is very likely that teams will respond to early signs of problems that threaten the planned progress and concentrate on solving them. When the habit of measuring performance is firmly established, the measures used can be more subjective and relate to less tangible and longer-term benefits, such as the number of innovations and new ideas generated.

There are two countervailing pressures to be resolved in developing effective sets of targets. First, the measures should reflect the specific objectives of individual teams. Second, they should be capable of being expressed in terms that make sense to the main boards of construction firms. Today, this increasingly means that the benchmarks will need to deal with more than just economic viability; most major firms accept the importance of also being environmentally sound and socially responsible.

Feedback

Feedback is absolutely crucial for construction to achieve improvements in its performance. It operates at every level and all teams should use systematic feedback to control their performance. This means organizations need to invest in developing robust feedback systems that can be used by all the teams involved and the teams should be trained in their use.

The information processed by control systems begins as raw data collected from the basic work teams about their progress and costs. It is analysed and summarized as it passes to other teams. The aim is to give each team feedback relevant to their own targets expressed in an appropriate language. For example, this means core teams require data describing the pattern of medium-term trends. This will tend to be fuzzy information which highlights risks and uncertainties. It will focus on patterns of performance and provide details of major sources of interference with planned progress or methods. At the level of basic work teams, information should relate directly to short-term performance. It will tell teams how their quality and safety record relates to the required standards, their chances of meeting the next milestone and their costs against budget. At this level, information will be clear and certain and is often best recorded in control charts that show trends in performance over time. In similar ways, all teams need feedback relevant to their own level of work.

It is especially important that strategic teams should make careful decisions about the feedback systems they use to guide their organization's work towards meeting agreed strategic objectives. Thus, if zero defects is a strategic objective, feedback on defects and their causes will be needed. If fast construction is a strategic objective, feedback on deviations from planned progress will be needed. These kinds of feedback should be produced regularly and discussed by the core and strategic teams as part of their ongoing search for continuous improvement.

To enable this to happen, strategic teams need to decide what aspects of individual projects and supply chains should be subject to feedback-driven control systems. Cost and time are not always the most important aspects; in many situations it is more sensible to treat quality and safety with greater rigour. Indeed, the great success of Japanese management methods is based on first controlling quality and then working to programme in a controlled manner. When quality and time are tightly controlled, efficiency is inherent in the work and cost control is virtually implicit. These issues should be debated by strategic teams in deciding what feedback systems to set up.

Strategic teams should consider how feedback can be used to guide new developments in their working methods. These may include developing the use of standards so that the nature of the organization's products gradually changes. For example, the use of prefabrication often begins experimentally in a series of trial approaches that are evaluated with the help of feedback. Improvements to processes also often depend on feedback. For example, improvements to the briefing process could link briefs to user feedback. Or improvements to construction planning could link programmes to the minutes of progress meetings. Or improvements to cost estimating could link initial estimates to final accounts. Significant gaps between any of these plans and actual performance should trigger a review of the processes involved.

Even when good feedback systems are in place it is important that executive teams do not forget the importance of direct, personal knowledge. They should 'walk the job' in the offices, factories and construction sites wherever basic work is taking place and ask questions such as:

- What are your targets?
- How are you doing against your targets?
- How well did you do last week?
- What is the next milestone you have to meet?
- What quality standards do you work to?
- Which other teams depend on your outputs?

Members of executive teams should work with an open door, in the sense that they will talk to anyone who comes to their office with a problem. On large projects, the equivalent of a suggestions box or a minister's red box is useful. This is a place where anyone can put a note, a copy of a document, a comment, a suggestion or anything they think the executive team ought to see. Such action does not remove the responsibility of subordinates to deal with problems they find, but it does provide a useful picture for teams responsible for the higher levels of work. Used in this way, red boxes supplement formal information and give members of the executive team clues about what to look for when walking the job.

This direct information, combined with systematic feedback data, provides the safest basis for making good decisions about construction work. Efficiency and innovation depend on both types of information being taken very seriously and being used by teams at all levels to guide continuous improvements.

Japanese control systems

The detailed planning that results in reliable quality in Japanese construction also ensures that projects always finish on time. This is possible partly because the programmes prepared at head office before work on site begins have slack built into them in various ways. When problems arise the slack is used to bring the work back onto programme. Thus, work is programmed on the basis of five normal working days each week; if necessary, work continues late into the night, seven days a week.

Each month's work is planned as four-week stages which leaves a few days each month to ensure that everything is exactly on target. When a problem is serious, a second shift may be worked; this may involve introducing a second subcontractor or, in an extreme case, a third. The aim is to keep the project exactly on programme or, when this proves impossible, to return it to the original programme as quickly as possible. The building must be completed exactly on the agreed date. It would be a matter of extreme dishonour and a great commercial embarrassment to any major contractor to complete a building even one day late; they would lose face and lose business. So everyone involved concentrates on doing whatever is necessary to complete all the required work exactly on time.

The importance of sticking to the plan is reinforced by the daily pattern of work. As a result, an important feature of the Japanese construction industry is that it is possible to describe a typical day on site because procedures are standardized throughout Japan. The working day begins with the whole site workforce lined up in subcontractor teams precisely at 8.00 am. The day begins with a fixed routine of stretching exercises learnt at school and practised daily throughout Japan. This ensures that everyone is warmed up and ready for work. Then the workforce is briefed on the planned day's work by the project manager. The briefing describes the main processes, major deliveries, safety priorities, quality issues and any other points which need special attention. Having prepared physically and mentally for the day's work, the subcontractor teams move to their workplace where each team holds what is called a toolbox meeting. This decides exactly how the team will achieve their agreed day's work. As a result, when work begins at about 8.25 am each worker knows exactly what he or she is to do during the day and where it fits into the overall plan. Each team continues its work until they have finished the agreed day's tasks.

Day-to-day coordination and control is provided by a meeting of the project management team and the foremen of all the subcontractors currently working on site. This is held at a fixed time each day, usually 3.00 pm, and any problems which have arisen during the current day's work are discussed and resolved. Then the next day's work is agreed in accordance with the overall programme prepared at head office before work started on site.

When problems are raised at the meeting, everyone focuses absolutely on finding a solution. All suggestions are considered in turn until a reasonable consensus is reached. Then the project manager announces the decision, which is accepted by all and recorded on a board in front of the meeting. These meetings take no more than 20 to 25 minutes and provide a clear and detailed resolution of any immediate problems and a plan for the next day. At the end of each meeting the project manager reads out the agreed decisions, highlights key points and reminds the foremen to be diligent in their work.

The most impressive feature of these meetings is that everyone concentrates on agreeing the best way to complete the work as originally planned. No-one challenges the programme or raises contractual issues, there is no mention of claims for extra money or time. Instead, there is tremendous peer pressure by the foremen on each other to maintain work exactly on programme and there is never any question of relaxing quality standards.

In contrast to the detailed control of quality and time, cost control in Japan is rudimentary. Agreed contract prices allow for producing a competent facility on time, including sorting out design, manufacturing and construction problems, and provides fair profit levels for contractors and subcontractors. Prices are set so as to force continuous improvements in performance. Customers expect to get better value in each successive project and so set tough prices. The plans made at head office before work on site begins are designed to achieve the customer's price and provide fair profits for all the construction firms involved, provided the quality and time targets are met. As a result cost control on site is little more than routine book-keeping to keep track of expenditure. Japanese managers act in the belief that if time and quality are controlled, cost will look after itself.

The incredibly simple approach to cost control works only because construction processes are planned in detail before they begin on site and subsequent effort is devoted to putting the plans into effect. It is true that Japanese construction teams also search for ways of improving their work but good ideas are reviewed, developed, tested and made robust before they are used, not on the current project but on future ones.

In the UK it is common for teams to look for better answers and apply them to their current project. This is why cost control has to be treated very differently. As Chapter 5 describes, the key features of best practice are that budgets are based on the value to the customer identified in the business case. Construction firms are paid an agreed profit and fixed overheads plus all their properly incurred direct costs in undertaking the work. Cost targets for the direct work are set on the basis that everything will go well. In other words, there are no contingencies built into the cost plans. Teams then search creatively for the best available answers. Cost control systematically identifies potential savings and cost threats and this is used to guide the core team, in cooperation with all the other teams involved, in doing everything possible to keep costs within budget. An important principle is that construction should be professional enough to accept all the costs that arise from the inherent uncertainties of construction. The price should be subject to change only if the customer changes the business case for the new facility. This should happen only very occasionally on mainstream projects but is likely to be more common on new-stream work.

Even with excellent cost control systems, costs cannot always be contained within agreed budgets. In Japan this is not allowed to distract teams on site from completing the project as efficiently as possible within the original plan. Then, if a contractor has indeed made a loss on a project for a regular customer, the outcome is discussed at a senior level between the two firms. If the contractor has finished on time, produced a good building and acted professionally in achieving this satisfactory outcome, the customer, conscious of their long relationship, will often accept that the price originally set was too ambitious and mitigate the loss either on the current project or on a future one. This is an important example of effective cooperation in action. Whilst in the long-run customers demand continuous improvements in performance, they feel a responsibility for ensuring that contractors stay profitably in business. The outcome in the UK is less certain and ranges from something close to the Japanese approach to expensive and time-consuming claims, arbitration or even litigation. Adjudication is increasingly used because it provides a reasonably cheap and quick way to resolve disputes over money. Best practice is to complete the work at the least cost, consistent with the quality standards and programme, irrespective of how any additional costs are dealt with. Then, when the project is finished and the full costs are known, decisions are made about what actions the construction organization should take to deal with any substantial cost overrun. Even though this conflicts with the provisions of many forms of contract used in the industry, it never-the-less provides the most reliable basis for building good relationships with customers and therefore for earning fair profits in the long run.

Meeting face-to-face

Procedures, standards, integral constraints and control systems are unlikely to provide all the information teams need. So discussions take place within and between teams. There are several effective ways of bringing people face-to-face.

The use of a team office to provide a common work place, especially where team members are drawn from several firms, has a number of benefits. Bennett and Jayes (1995, 1998) describe examples in which a common team office encouraged a wide involvement in key decisions. One good approach is for all meetings to take place in the team office. Decisions are recorded on a board that is left on open display until everyone has had a chance to consider what has been decided, ask questions and contribute further ideas. Underlying this approach is the principle that everyone has the right to contribute to team decisions because they have a responsibility for putting them into effect.

Team offices can allow almost subliminal flows of information as people half-hear conversations, half-see drawings and programmes, absorb an understanding of financial problems and sense the real state of progress. This background knowledge can and often does trigger a warning signal when the first faint signs of a problem surface and when there is still time to deal with it before it becomes a crisis.

As a result, team offices are often used for work that requires a lot of new information to be exchanged. They help ensure that decisions are more thoroughly considered and made faster. Case studies suggest that the early creative stages of work benefit most from team offices. The subsequent, more routine work is often best done in firm's own offices where people have fewer interruptions and good access to specialized information and support.

Workshops are another important way of bringing people together. As described earlier in this chapter, workshops provide important benefits when key work stages are reached. They help build cooperative behaviour between teams and so are commonly held when teams start new work of any kind. The early days of a new project are frequently marked by a formal workshop; then as the work progresses, further workshops are held as often as there are major decisions needing wide-ranging discussion. A final workshop is an important part of feedback systems since it gives an opportunity for lessons to be identified, reconsidered and recorded.

Social events also play an important part in building team spirit and helping people communicate effectively. Many effective teams plan regular social events that team members are expected to attend. In addition they ensure there are regular opportunities to meet outside of formal work situations in a relaxed atmosphere. This builds up a rounded picture of colleagues which helps when difficult decisions have to be made.

In addition to these various ways of bringing people together, most work requires a formal structure of meetings to agree plans, ensure progress, solve problems and make decisions. A typical pattern of meetings for a new-stream project is given in *Table 6.1*. Meetings inevitably take up a lot of time for many people and so it is important that they are well run. The following matters should be taken into account.

Meetings should always have a clearly defined purpose. Everyone with an interest in the outcome should be present or represented in some way. There should be a nominated chairperson who begins by making sure that everyone knows one another; if there is a new person present, they are introduced and everyone else introduces themselves. Next the subject matter is defined by checking what everyone wants to achieve from the meeting. The meeting should then agree a clear agenda that allocates sufficient time for decisions to be properly considered, and settles the finishing time so that discussions are purposeful. Decisions and actions should be recorded and agreed at the meeting so there is no need to circulate paperwork. When all the agenda items have been dealt with, the chairperson should close by checking how everyone feels about the meeting and whether they have achieved what they wanted.

Crises and problems

An important part of the management tool of meeting face-to-face is having well rehearsed ways of dealing with crises and problems. Resources should be focused on finding permanent answers in order to facilitate a return to normal planned and controlled working as soon as a good answer has been found.

Table 6.1: Typical pattern of meetings for a new-stream project

Team meetings
All teams meet as often as they need to complete their work. In addition the following meetings involve more than one team

Milestone meetings
The core and executive teams meet as each milestone is reached (typically one a month) to ensure that the current milestone has been met, to review performance and plan the work leading to the next milestone

Substage meetings
The core and executive teams meet every week to solve any problems not being dealt with elsewhere. They review the design, working methods, progress and cost reports and external influences to ensure the project is meeting agreed targets and make decisions about problems and opportunities

Design meetings
The executive team and key members of the technology cluster teams currently involved in design meet every week to review the design, any feedback from the customer and supply chains, explore alternative solutions, ensure that the design is meeting all the agreed targets, and resolve any design problems. Once all the design decisions are made, these meetings cease

Technology cluster meetings
Each technology cluster team, with the key members of the direct work teams currently involved in the work, meet every week to ensure that all necessary information is in place for the coordination of the design and for efficient construction, to look for better ways of working and resolve any problems. These meetings take place as long as work on the technology cluster is still under way

Direct work meetings
Key members of the executive team meet every week with each specialist contractor's contract manager and site supervisor to resolve any design and procurement problems, agree the exact scope of the work, agree the terms for it to be carried out including ensuring there is a fair basis for payment, and check that the work meets all the agreed targets. Once these decisions are made, subsequent meetings concentrate on ensuring that detail design, construction method statements and construction are meeting all the agreed targets, and resolve any problems. The meetings are held as long as the specialist contractor is working on the project

Construction meetings
Daily meeting of the executive team and site supervisors currently working on site to review the day's progress; solve any problems affecting the day's progress; and plan the next day's work in detail. These meetings are held right through the construction stage

The best way of dealing with crises and problems is to anticipate them and get teams to practice and rehearse solutions. Then, when real crises or problems arise, less disruption is caused. Also rehearsal often means that a ready-made answer already exists.

The key principle in dealing with crises and problems that threaten the agreed programme is to maintain milestones but review what must be done by the due date in the light of actual progress. For example, ideally when the milestone for beginning the fabrication of structural steelwork is reached, the steel fabricator will have a complete design for every steelwork component and all the connection details. In practice, as Bennett and Jayes (1998) report in a case study of a very successful building project for Rover, steel fabrication can begin on the basis of much less information. The approach used for this construction work was based on that used for designing and producing cars in Rover's joint venture with Honda. Thus, it was based on Japanese ideas overlain with a UK desire to keep searching for the best answers right up to the last possible moment for making a decision. Perhaps inevitably, when the steel fabrication milestone was reached, the core and executive teams faced a crisis because a number of key decisions had not been made and there was a major risk that the project would be severely delayed. They brainstormed the situation and decided to authorize the steelwork contractor to begin fabricating the steel components on the basis of agreed design assumptions. There were risks involved in this and indeed some of the assumptions had to be changed when the design was worked out in detail. As a result, the steelwork contractor had to alter steel components already manufactured and extra costs were incurred. However, the executive team's approach meant that the steel frame milestone was met. By working flexibly in this way within fixed milestones, the overall programme was achieved.

The implications when crises or problems threaten the agreed costs are different. The key requirement is for everyone involved to accept responsibility for staying within cost targets. Detailed cost feedback that identifies cost threats and opportunities for savings is needed. This should be reviewed by all the teams at intervals that reflect their level of work. For example, core and executive teams normally look at cost reports every week and review all the threats and opportunities identified. These reviews should lead to decisions that deal with cost threats and adopt sufficient of the savings to ensure that work stays within the overall budget. Thus in the Rover case study mentioned above, the additional costs involved in altering steel components were identified as a cost threat. It was then the responsibility of the whole team, not just those responsible for the steelwork, to join in the search for savings to offset the extra costs. This was, in fact, achieved and the building completed to Rover's satisfaction within the original budget.

A fundamental principle implicit in these approaches is that people are encouraged to look at their own contributions to problems and not try to blame others. Solutions come when everyone involved concentrates on what they can do to help solve the problem, not worries about what others should do. Focusing on what other people should do merely creates conflicts and implies blame.

Workshops are an especially important part of good problem-solving procedures because they allow different interests to be considered and reconciled. They give everyone involved the confidence that their interests have been taken into account. Even when a party does not get everything they wanted, the sense of fairness that comes from a well run workshop helps them accept that they have got as much as is reasonable.

Workshops often do more than this and produce answers that enable everyone to have all they really need and more. Win:win solutions can be found when people are open about their real interests and cooperate in looking for the best answers. So workshops should be used throughout projects to deal with major crises and problems.

Difficult problems may justify setting up a task force to find a good answer. As Chapter 5 describes, a task force is a group of individuals drawn from several teams brought together specifically to solve a difficult problem. Task forces work best when they are set up to solve one problem and are given a relatively short time to find an answer, and are disbanded when their work is done. The members should be experienced in the subject of the problem so they will understand its causes and possible answers.

In looking at all ways of tackling crises and problems it is good to have a wide perspective. Solutions often come from looking at problems from different points of view. Ask: How does this compare with other problems? How would it be tackled by other people? Other people could include, for example, an architect, an engineer, a craftsman, a local politician, a television reporter, a doctor or a customer. It helps to actively look for the positive aspects of any situation, no matter how bad it may appear. People choose how they respond to situations, so whether they see a given set of circumstances as an opportunity or a problem is a matter of choice. Whether any given situation is good or bad is a matter of how people view it; all have some good and some bad features, people choose which they concentrate on. It is much more efficient to concentrate on the good. If some situation really is bad and people can do something about it or learn some lessons from it, they should do so and then move on to the next task. If there really is nothing they can do about the situation, they should just accept the inevitable and not waste time and energy dwelling on it.

Particular problems arise when an individual in a team acts selfishly against the interests of the rest of the team. Strong counter-measures must be taken but not on the basis of blame, anger or hate. The first step should be a clear discussion of the situation and the interests involved. Then actions should be taken that are commensurate with the selfish acts. These may include detailed checking and additional audits of the person's work, a requirement for frequent progress reports or other similar routines that focus on the team's interests. If the selfish acts are repeated, the actions should be strengthened and made more urgent. If they still continue, the individual should be replaced on the basis that everyone deserves a second chance but not an endless stream of chances. These procedures should be clear and certain so that no-one believes that they can exploit the team for their own advantage and get away with it.

When a team is unable to find an answer to a crisis or problem that enables them to keep their work within its defined constraints and targets, there should be a procedure for referring it to another team with a wider responsibility. Thus a work team should try to solve crises and problems but, if they cannot find an answer quickly, say within two working days, they refer it to the executive team. If they are unable to find an answer within a further two days it is automatically referred to the core team to be resolved in, say, three days. The common effect of such procedures is that the vast majority of problems are resolved quickly by work teams.

It is when teams face crises and problems that they are most in need of good information systems. It may well be that good answers already exist elsewhere or, at least, other teams have faced similar situations. A good information system will allow the crisis or problem situation to be described on an intranet and so identify people with directly relevant answers or experience. If that fails, the system will look for ideas that may provide clues to an answer. In effect the system enables the team to consult the collective memory and experience of the whole organization and all its contacts with other organizations. Information systems are more than just a source of solutions to problems, they are crucial to effective work.

Information systems

Information systems provide organizations' formal memories. They consist of data bases and formal flows of information inside an organization. The overall aims of information systems are to identify relevant information and distribute it to everyone who needs to know, in a convenient form, at the appropriate time. It is obviously essential for information about objectives and design and management decisions to be widely distributed. Teams need to know what they are required to do in sufficient time to plan and organize their work. They also need to know how their work relates to that of other teams.

The scope and effectiveness of information systems depends on the time and effort spent developing them and the nature of the work to which they relate. Mainstream information systems should be well developed and comprehensive. When they are based on continuity that has persisted over many years, the systems can provide impersonal and virtually automatic control of work. This level of control does not as yet exist in construction but can be seen, for example, in the operation of oil refineries, chemical processing plants and automated factories. Never-the-less, aspects of construction are moving in this direction and computer-aided design systems linked to expert systems that provide construction management information are already routinely used in the design of some mainstream facilities.

New-stream information systems tend to be comparatively rudimentary and provide little support for teams. In part this is due to individualistic professionals denigrating the use of standard answers and claiming that procedures inhibit creativity. This is an error because construction has well developed industry standards and procedures that apply to nearly all projects. It is an error that the industry needs to put right because modern information technologies make it possible for teams to have available huge amounts of detailed information whenever it is needed.

In many other industries, modern information technologies provide the basis for sophisticated information systems that link analytical tools to information so that the systems make routine decisions. These developments lead to automated production systems which are making industry after industry totally dependent on their information systems.

Firms in many leading manufacturing industries now rely on individual customer's orders to pull the required components and subcomponents through their supply chains. Suppliers and subcontractors manufacture and supply exactly what is needed to meet customers' orders and the parts all come together in a predetermined sequence for final assembly. At the leading edge of manufacturing this all happens with little human intervention because highly developed information systems drive the processes virtually automatically.

In retailing, information systems now enable sales to pull new stock through stores, warehouses, distribution systems, packaging plants and production facilities. Each sale helps trigger a stream of decisions throughout these integrated, widely distributed systems. The same systems provide information about customers' interests and preferences to guide future marketing strategies.

The same technologies are blurring the boundaries between industries. Supermarkets offer banking, insurance, finance and other information-related services at the check-out points where customers do their weekly shopping. The convergence of powerful information and communication technologies is the enabling force in creating these new businesses.

Already many products and services are marketed through the Internet so potential customers can browse through suppliers' information, reviews, market surveys, test results, consumer evaluations and related information. This information is organized so it can be accessed in text, graphics, sound, video or whatever form the customer finds most convenient.

The most dramatic effects of the new information and communication technologies are in creating businesses that are truly global. 'Think global, act local' is a widely quoted slogan that reminds multi-national firms to plan and organize production on a global scale but adapt to local needs and fashions in marketing their products. This is a long way from being a reality in most industries. There are, however, exceptions notably in the provision of financial and marketing services where information is the product. Firms in these industries can operate globally because their information systems are global. They deliver massive amounts of information linked to powerful analytical tools to wherever decisions need to be made. Products and services can be tailored to the needs of precisely defined categories of customers and marketing strategies designed to fit local situations.

In all industries information is the key to providing customers with better products and services, convincing them that they are indeed getting good value, and delivering what is promised quickly, efficiently and reliably. This is as true for construction as for any other industry, yet construction has been slow to use the potential offered by integrated information and communication technologies. This is not because of limitations in the technology; construction's problem is that it has not standardized its information sufficiently to make use of the technologies driving other leading industries. Once construction teams have established what information relevant to their work needs to be available, they can be confident that technologies exist to turn it into effective information systems.

TEAM INFORMATION SYSTEM
Products and services
Standards
Processes
Procedures
Constraints
Targets
Benchmarks
Control sytems
Feedback
Problem resolution support

Figure 6.11: Checklist for team information systems

This chapter provides a guide to issues that need to be considered by teams designing their information systems. *Figure 6.11* provides a checklist of the main categories of information needed by teams. It is particularly important to give teams access to wide-ranging information that may trigger new ideas to help them resolve problems. Access to well organized information sources is a key part of finding new answers and so of teams improving their performance. The Internet promises to provide the basis for this and construction should be using this modern wonder of the world to support teams throughout the industry.

Long-term development

Once the framework of procedures, standards, constraints, control systems, meetings and information systems shown in *Figure 6.1*, is in place, teams have a realistic chance of working effectively. However, this should be seen merely as a starting point, a datum from which they can begin to concentrate on becoming more efficient. Construction's future depends on teams searching relentlessly for better ways of working. Delivering better value and quality in ways that provide real benefits for customers provides many benefits, not least that the demand for construction will be higher and more consistent.

In addition to the ubiquitous need for continuous improvement, it is inevitable that markets and technologies will demand further changes. It is in construction's best interests to guide these developments, or at least be sufficiently involved to be aware of what is happening so as to be ready to meet new demands.

Feedback should be a key driving force for the continuous improvement of all construction's processes and for more fundamental long-term developments. This means gathering information about markets and technologies from all parts of construction organizations, especially from teams involved in individual projects. It means gathering feedback from customers about new ways of using their constructed facilities. It means monitoring distinct market sectors, relevant R&D, new ideas in current practice, new products and services emerging in supply chains and new initiatives by competitors. It means being involved in major debates about political, economic, social and technological change. It means understanding changes in the cultures, values and economies of national and international communities. It means anticipating and understanding change.

These wide-ranging information gathering systems help make construction at all levels responsive to current trends. This may mean setting new kinds of objectives for project teams. It may well mean setting up task forces to explore potential new developments. It may mean commissioning new-stream projects to explore a new market opportunity or to try out new technologies. It may mean investing in developing a supply chain. It may mean bringing new firms into the industry because they provide skills and knowledge likely to become important. It may mean devising new facilities management services to meet changing needs. It may mean finding new ways to finance construction.

In choosing amongst these actions, construction should plan and invest in ways that provide continuity for teams. This is the key that makes it sensible to invest in the future. It makes it sensible to invest in training so that workers are using the most efficient methods. It makes it sensible to invest in dedicated plant and equipment so that manual work is reduced. It makes it sensible to invest in research and development to find better answers. In other words, ensuring continuity for teams provides a basis for construction to behave as other manufacturing industries and invest in the quality of its products and services and the competence of its workforce.

Standards and procedures provide a datum for improvement. They should reflect current best practice and, to ensure that this remains the case, they should be reviewed at regular intervals so that any which are not serving the organization's long-term objectives can be pruned out.

When problems arise, they should always be taken seriously and regarded as opportunities for finding permanent improvements. All the teams responsible for work where a problem has arisen should focus their attention on searching for a long-term answer. Problems should be tackled immediately, when and where they occur. In this way the source and causes of each problem are likely to be identified accurately. Any temptation to blame individuals should be resisted because virtually all problems have systematic causes. Therefore it is everyone's responsibility to help identify the fundamental causes of problems and find permanent solutions.

For all this to happen managers must relinquish control and instead concentrate on empowering workers at all levels. This may well mean that workers have to be trained in process analysis, work measurement and statistical control techniques so that they have the tools to improve their own work. Managers should be active in encouraging workers to search for a steady stream of well thought-out innovations to improve both their products and processes. The absolute importance of reliable quality in every aspect of work needs to be constantly reinforced by providing excellent working conditions, encouraging workers to maintain a smart appearance, to wear clothes appropriate to the work in hand, and to keep their workplace clean and free of rubbish and waste material. They should be encouraged to set ever-tougher targets for themselves, and to question everything about their work in the search for better answers.

The overall strategy for successful long-term development is first to achieve better control and greater certainty in performance, and then to search for improvements. For mainstream organizations this means the continuous improvement of a well developed approach embodied in procedures and standards. Decisions that shape these long-term processes should be made by representatives of all the firms involved in the overall construction organization, including its customers and supply chains, and the communities affected by the new developments. The search for ways of improving existing answers should be relentless until the rate of improvement slows or information about markets and technologies suggests that the approach will soon become out-dated. At this point mainstream organizations need to make a step change.

New-stream organizations are engaged in a constant search for new ideas. As a result some of their work provides a basis for mainstream step changes. This link allows both mainstream and new-stream to concentrate on what they do best and between them provide customers with ever-improving value.

The overall approach balances cooperation with beneficial competition that provides a spur for continuous improvement. This balance is not easy to achieve, as Japan's current problems of over-capacity in outdated answers testify. Japan has continued too long with improving mainstream approaches when changing markets and technologies required step changes. The new paradigm helps avoid this weakness by making processes responsive to wide-ranging feedback, and providing a pool of new answers ready to replace current technologies as soon as it is apparent they are no longer competitive. Thus, the third way steers a course between an over-reliance on cut-throat competition and a too cozy reliance on cooperation. It does this, not by making comfortable compromises, but by tough-minded decisions made by and in the interests of all the people involved in and affected by construction.

References

Atkins, W.S. and consultants including Centre for Strategic Studies in Construction, The University of Reading (1994), *Strategies for the European Construction Sector.* Luxembourg: European Communities.

Axelrod, R. (1984), *The Evolution of Cooperation.* New York: Basic Books.

Beer, S. (1972), *Brain of the Firm.* Harmondsworth: Penguin.

Belbin, R.M. (1993), *Team Roles at Work.* Oxford: Butterworth-Heinemann.

Bennett, J. (1991), *International Construction Project Management.* Oxford: Butterworth-Heinemann (A revised and summarized version is available at www.architecturalpress.com).

Bennett, J. and Jayes, S.L. (1995), *Trusting the Team: The Best Practice Guide to Partnering in Construction.* London: Thomas Telford.

Bennett, J. and Jayes, S.L. (1998), *The Seven Pillars of Partnering: A Guide to Second Generation Partnering.* London: Thomas Telford.

Bennett, J., Flanagan, R. and Norman, G. (1987), *Capital and Counties Report. Japanese Construction Industry.* Reading: Centre for Strategic Studies in Construction.

Bennett, J., Pothecary, E. and Robinson, G. (1996), *Designing and Building a World-class Industry.* Reading: Centre for Strategic Studies in Construction.

Blair, T. (1998), *The Third Way: New Politics for the New Century.* London: Fabian Pamphlet 588.

Broadbent, G. (1973), *Design in Architecture.* Chichester: Wiley.

Capra, F. (1996), *The Web of Life.* London: Harper Collins.

Carlisle, J.A. and Parker, R.C. (1989), *Beyond Negotiation: Redeeming Customer–Supplier Relationships.* Chichester: Wiley.

Chandler, A.D. (1977), *The Visible Hand: The Managerial Revolution in American Business.* Harvard: Harvard University Press.

Construction Industry Board Working Group 12 (1997), *Partnering in the Team.* London: Thomas Telford.

Construction Industry Institute (1991), *In Search of Partnering Excellence.* Austin, TX: Bureau of Engineering Research, The University of Texas in Austin.

Construction Task Force (1998), *Rethinking Construction*. London: HMSO.

Cooper, R. (1996), *The Post-Modern State and the World Order*. London: Demos.

Covey, S.R. (1989), *The Seven Habits of Highly Effective People*. New York: Simon and Schuster.

Dawkins, R. (1986), *The Blind Watchmaker: Why the Evidence of Evolution Reveals a Universe Without Design*. New York: Norton.

De Geus, A. (1995), Companies: what are they? *RSA Journal*, 5460: 26–35.

DTI Overseas Science and Technology Expert Mission to Japan (1995), *Time for Real Improvement: Learning from Best Practice in Japanese Construction R&D*. Ascot: The Chartered Institute of Building.

Duffy, F. and Henney, A. (1989), *The Changing City*. London: Bulstrode Press.

Fisher, R. and Ury, W. (1981), *Getting to Yes*. London: Hutchinson Business.

Flanagan, R., Ingram, I. and Marsh, L. (1998), *A Bridge to the Future: Profitable Construction for Tomorrow's Industry and its Customers*. London: Thomas Telford.

Galbraith, J. (1973), *Designing Complex Organizations*. Reading, MA: Addison-Wesley.

Giddens, A. (1998), *The Third Way: The Renewal of Social Democracy*. Oxford: Blackwell.

Gray, C. (1996), *Value for Money: Helping the UK Afford the Buildings it Likes*. Reading: Reading Construction Forum.

Gray, C. and Flanagan, R. (1989), *The Changing Role of Specialist and Trade Contractors*. Ascot: Chartered Institute of Building.

Gray, C., Hughes, W. and Bennett, J. (1994), *The Successful Management of Design: A Handbook of Building Design Management*. Reading: Centre for Strategic Studies in Construction.

Green, S.D. (1994), Beyond value engineering: SMART value management for building projects. *International Journal of Project Management*, 12, 49–56.

Handy, C. (1994), *The Empty Raincoat: Making Sense of the Future*. London: Hutchinson.

Heirs, B. and Pehrson, G. (1982), *The Mind of the Organization*. New York: Harper & Row.

Henderson, H. (1993), *Paradigms in Progress: Life beyond Economics*. London: Adamantine.

Hillebrandt, P.M. (1984), *Analysis of the British Construction Industry*. London: Macmillan.

Hordyk, M. and Bennett, J. (1989), *New Steel Work Way*. Ascot: The Steel Construction Institute.

Hutton, W. (1995), *The State We're In*. London: Cape.

Imai, M. (1986), *Kaizen: The Key to Japan's Competitive Success*. New York: McGraw-Hill.

Jaques, E. (1989), *Requisite Organization: The CEO's Guide to Creative Structure and Leadership*. Arlington, VA: Casson Hall.

Kanter, R.M. (1989), *When Giants Learn to Dance: Mastering the Challenges of Strategy, Management and Careers in the 1990s*. London: Simon and Schuster.

Kidder, R. (1999), Global ethics and individual responsibility. *RSA Journal*, 1. 1999, 38–43.

Kuhn, T.S. (1962), *The Structure of Scientific Revolutions*. Chicago, IL: University of Chicago Press.

Lancaster, C.L. (1994), *The J6 Partnering Case Study*. Washington, DC: US Corps of Engineers.

Latham, Sir M. (1994), *Constructing the Team*. London: HMSO.

Lazonick, W. (1991), *Business Organization and the Myth of the Market Economy*. Cambridge: Cambridge University Press.

Locke, R.R. (1996), *The Collapse of the American Management Mystique*. Oxford: Oxford University Press.

Luttwak, E. (1998), *Turbo-capitalism: Winners and Losers in the Global Economy*. London: Weidenfeld & Nicolson.

Lynton (1993), *The UK Construction Challenge*. London: Lynton.

Maslow, A.H. (1954), *Motivation and Personality*. New York: Harper & Row.

Morgan, G. (1986), *Images of Organization*. London: Sage.

Oliver, N. and Wilkinson, B. (1992), *The Japanization of British Industry. New Developments in the 1990s*. Oxford: Blackwell Business.

Open Systems Group (1981), *Systems Behaviour*. London: Harper & Row.

Peters, T. (1987), *Thriving on Chaos*. London: Macmillan.

Peters, T.J. and Waterman, R.H. (1982), *In Search of Excellence*. New York: Harper & Row.

Pickrell, S., Garnett, N. and Baldwin, J. (1997), *Measuring Up: A Practical Guide to Benchmarking in Construction*. London: Construction Research Communications.

RSA (Royal Society for the Encouragement of Arts, Manufactures and Commerce) (1995), *Tomorrow's Company: The Role of Business in a Changing World*. London: RSA.

Schon, D.A. (1983), *The Reflective Practitioner*. London: Temple Smith.

Smith, A. (1776), *The Wealth of Nations*. London: Stratton and Cadell.

The Economist (1999), A survey of business and the Internet: the net imperative. 26 June 1999

Vermande, H.M. and Van Mulligen, P-H. (1999), Construction costs in the Netherlands in an international context. *Construction Management and Economics*, 17: 3.

Womack, J.P. and Jones, D.T. (1996), *Lean Thinking: Banish Waste and Create Wealth in your Corporation*. New York: Simon and Schuster.

Womack, J.P., Jones, D.T. and Roos, D. (1990), *The Machine that 'Changed the World'*. New York: Macmillan.

Index